Ch[illegible]p [illegible]

Intel v. AMD ; Intel v. IBM

BRIAN W. KELLY

Copyright © 2005, Brian W. Kelly
Chip Wars Author Brian W. Kelly

All rights reserved: No part of this book may be reproduced or transmitted in any form, or by any means, electronic or mechanical, including photocopying, recording, scanning, faxing, or by any information storage and retrieval system, without permission from the publisher, LETS GO PUBLISH, in writing.

Disclaimer: Though judicious care was taken throughout the writing and the publication of this work that the information contained herein is accurate, there is no expressed or implied warranty that all information in this book is 100% correct. Therefore, neither LETS GO PUBLISH, nor the author accepts liability for any use of this work.

Trademarks: A number of products and names referenced in this book are trade names and trademarks of their respective companies. For example, NetBurst is a trademarks of Intel Corporation; MirrorBit is a trademark of AMD Corporation and Power Everywhere is a trademark of IBM Corporation. A list of a number of trademarks is included following the front matter.

Referenced Material : The information in this book has been obtained through personal and third party observations, interviews, and copious research and analysis. Where unique information has been provided or extracted from other sources, those sources are acknowledged within the text of the book itself. Thus, there are no formal footnotes nor is there a bibliography section. Any picture that does not have a source was taken from various sites on the Internet with no credit attached. If any picture owner would like credit in the next printing, please email the publisher.

Published by: LETS GO PUBLISH!
Joseph McDonald, Publisher
P.O Box 425
Scranton, PA 18503
jmac@letsgopublish.com
www.letsgopublish.com

Library of Congress Copyright Information Pending

Book Cover Design by Michele Thomas

ISBN Information: The International Standard Book Number (ISBN) is a unique machine-readable identification number, which marks any book unmistakably. The ISBN is the clear standard in the book industry. 159 countries and territories are officially ISBN members. The Official ISBN For this book is:
0
0-9745852-5-4

The price for this work is : **$29.95 USD**

10 9 8 7 6 5 4 3 2 1

Release Date: **August, 2005**

Trademarks

Intel, Itanium, Pentium, Xeon, NetBurst are registered trademarks of Intel Corporation or its subsidiaries in the United States and other countries.

AMD, Athlon and Opteron, MirrorBit, are registered trademarks of AMD Corporation.

IBM, iSeries, pSeries, eServer, PowerPC, POWER5, System/3, System/32, System/36, System/38, AS/400, .and Cell are registered trademarks of IBM Corporation

Microsoft, Windows, Windows XP, and Windows NT are registered trademarks of Microsoft Corporation

Note: Any vendor wishing that a trademark citation be placed in a subsequent printing of this book is asked to email the publisher.

Front cover picture courtesy of IBM

Chips on back cover courtesy of AMD and Intel

Dedication

To my wonderful wife Pat, and our loving children, Brian, Michael, and Katie as well as the greatest pack of brothers and sisters, aunts and uncles, cousins, nieces and nephews that any person could ever hope to have in just one life.

Acknowledgments

I would like to thank many people for helping me in this effort.

I would first like to thank my immediate family, starting with my lovely and dear wife, Patricia. Again, as I offer in all my books, my wife Patricia is my source. She is the person who keeps me alive and sane and well in more ways than can be mentioned. She is the glue that holds our whole family together. Besides that, she keeps getting better looking as the years go by, and I love to see her wonderful face every day in my life. Her daddy, a wonderful man in his own right, Smokey Piotroski, called his little girl *Packy* as a nickname. Though Stash is now with the Angels, I love that name and the person who wears it and I still use it to address my little Packy. God gave me a gift that keeps me going. Thank you Packy for all you do to keep me and our whole family well and mostly, thank you for the smile that you always put on my face.

I would also like to thank my twenty-year-old daughter, Katie, who is still my little baby doll. Kate helps me in any way she can. Even more than that, her sweet voice and her accomplished guitar playing gets the muse racing as my fingers pound the keyboard. Katie is starting to feel better now and we thank God for that but it still is not easy for her. She is on her way to being OK. I thank my Katie for she will always be Daddy's Little Girl. I love you very much. A special thank you also goes to Dr. Patrick Kerrigan, who is working his way up the list in record time. Dr. Patrick comes to the job with the abilities of Hippocrates and the patience of Job. He has left no stone unturned in helping Katie through her illness.

Thanks also go out to my twenty-three-year-old son, Michael, who more than made the deans list in his last semester at King's College this year as he received his B.S. Degree in Accounting. Michael had very good LSATs and his academic record was more than enough for him to be accepted in Law School and he will be beginning his three year trek this month. I am always happy to have my youngest son close by so I am going to miss him very much.

I also thank my twenty-four-year-old son, Brian, who just knocked 'em dead in Law School and graduated Magna Cum Laude. Brian spent most of this summer taking courses in preparation for the Bar Examination, which he took last week. He thinks that he did very well and that he'll be getting permission to practice soon. After a short break, Brian will be starting a clerkship. Congratulations

Brian, you make us very proud. Mom and I are very proud of all of our children and we thank each of them for their work in academia and their efforts on our behalf.

Accomplishments often materialize because of a strong friendship infrastructure. I am pleased to have a number of great friends. Among them is my longtime best friend Dennis Grimes, who is always there to help, though he may think everything I write sounds the same. Professor Grimes is on the faculty with me at Marywood University and he is a CIO for Klein Wholesale. He is very talented and very helpful. I selected his comments about Chip Wars as a cover quote. Barbara Grimes, Patricia Grimes Yencha, Elizabeth (Wizzler), Mary the PhD., Denyse back from the U.K., Grandma Viola, and Grandma Gert also pitch in whenever the opportunity arises. Dennis helps me in whatever way I ask, especially when I am stuck. I really appreciate all you do for me "D." Thank you

The busiest guy on all of my book projects, besides myself, is always Joe McDonald. Joe is the businessman in our publishing venture, and in that, he's all business. Joe is the former Publisher of the Scranton Tribune/Scrantonian Newspaper. So he's got the right background to make sure everything is A-OK! I promised Joe that my next book was going to be non-technical as we moved the publishing business to Children's books and third party authors. Joe assures me that after this book, he will have the courage to lead me to the children's side of the business where our next book is scheduled to be The Adventures of Eddy (The Dog) written by Joe's Grandson. Soon, it will be on the bookshelves of America. My thanks also go to Peg McDonald for making sure that Joe is always ready for action.

Of course, the long list of helping hands contains lots of names: Gerry Rodski, Jeanne and Farmer Joe Elinsky, John and Carol Anstett, Grandma Leona and Grandma Annie (from Mayflower), Carolyn and Joe Langan, Bob & Cathy Wood, Karen and Al Komorek, Bonnie and George Mohanco, Becker and Robin Mohanco, Lilya, Josh, and Alaina Like Mohanco, Bob and Nettie Lussi, Kim and Ruth Borland--- they are all there when needed.. Other helping hands include Dr. Lou and Marie Guarnieri as well as Mary and Cindy Guarnieri, whose hands have been indispensable. I can't forget Frannie and Mike Kurilla, Jerry and Hedy Cybulski, Linda DeBoo and Bob Buynak. Joe, the Chief, LaSarge, John and Susan Rose, and Dave and Nancy Books. Special thanks also to the E.L. Meyers Class of 1965 (40th reunion this year) for some early training in the art of writing.

And don't let me forget Patricia's parents, Arline and Stanley Piotroski, who continue to guide us in our lives. Cathy and Marty Piotroski, Dr. Susan Piotroski and Dr. Mitch Bornstein, Matt and Allie, Dr. Stan Piotroski, Carol Piotroski, Sister Marlene, Justin and Katie, Merek, MacKenzie, Myranda, Erin, Ralph Harvey, Lynn, and Scott Piotroski, Pierre Le Kep. The Kelly parents -- Ed and Irene also provide guidance from upstairs as well as direct intervention as needed; Anna Maye, Nancy and Angel Jim Flannery (Leland (No K) Zard), Renee (Bean), Jimmy (Jim Bob), Bridget, Mary (MeeWee), Danny, Michael (McPike) , Ken (La Rue), Jen, Angel David Davidow (Brunoch Zard), Stephen (P.Q. Whoozer), Matthew(M.Q. Peph), Bailee Roo, Viva La Vieve, and Billiard Peph, Joe and Rosalee, Raymond and the real Sparkey. Mary and Bill Daniels, Liz (Weezler), Bri, Meg (Gledeebaigledee), Bill Jr (Billdog) ., Vicky, Sophia (Chubby Cheeks,) Elise (La Leese), Diane and Joe Kelly, Tara and Col, Ed and Eudart Kelly, Eddie, John, and Robert. Barb, Kim & Dave, Dawn, Cindy & Dave Boyle. Bill Rolland- Notre Dame's # 1 Fan and master of accommodations, Bill Kustas, Bill & Helen Kush, Steve and Shelly Bartolomei.

Going back to the top of the list of helpers is my wonderful and huge pack of cousins. The list begins with the Uncles and Aunts, many of whom are now Angels. Uncle Nick and Aunt Emma McKeown, Dave and Kathleen Conklin, Rita and Frank DeRiancho, Joan and Tom Nelson, Aunt Ruth and Uncle Joe McKeown, Kathy and Joe McKeown Jr., Aunt Louise and Uncle Jimmy McKeown, Patsy, Danny and Jerry McKeown, Nina and Jim Brady, Jimmy Brady, Tommy and Mary Rowan, Arlene and Richard May, Little Tommy Rowan, Helen and Joe Drexinger, and all the other cousins, uncles & aunts who can't make it to the special muse event every summer in Montrose.

Of course, there's Uncle Johnny Kelly, Aunt Catherine and Leonard Lamascola, Aunt Mary Kelly, Sharon, Maureen, Jud, Pat Jr., and Tommy Kelly. Red Cloud is also on the list for his due diligence in writing postcards.

In the special care category, Dr. Lou has been making sure that my bones are aligned properly for years. Dawn Acacio has recently begun to apply her PT trade to my sore neck so I can keep squeezing out paragraphs. And, so that I can give those speeches with a bright smile, I get some big help from Dr. Lou Kicha the Great and his highly competent team of professionals at Aspen Dental-- John Cicon, Carol Kephart, Nicole Arnone, Anita Florek, and the tooth architect, Mary Lou Lennox. Thank you all very much.

Special acknowledgments to Steven Dressler and Howard Klein, the top management team at Klein Wholesale Distributors in Wilkes-Barre, PA, who use chip technology to the fullest. Their vision, foresight, and execution have brought Klein to the enviable position of being the third largest candy and tobacco wholesaler in the United States.

Various members of the Klein development staff offered information over the time in which this book was written. In alphabetical order, by first name, the Klein team includes: Barb Chaderton, Bill 'Curly' Kepics, Cindy Dorzinsky, Cindy Goodwin, Dennis Grimes, Eric Priest, Jeff Massaker, Jerry Reisch, Joe Byorick, Joe Rydzewski, John Robbins, Paula Terpak, Rod Smith, and Rosalind Robertson.

I would also like to thank Nancy Lavan, our sponsor at Offset Paperback, our printer. She continually encourages us in our writing and publishing efforts. Chris Grieves, our new customer service person has made working with the printing process an easy task. Special thanks go to Michele Thomas, who takes ideas and makes wonderful images from them, such as this wonderful cover.

To sum up my acknowledgments, as I do in every book that I have written, I am compelled to offer that I am truly convinced that "the only thing you can do alone in life is fail." Thanks to my family, good friends, and a helping team, I was not alone.

Table of Contents at a Glance

Table of Contents

Chapter 10 Challenger: Advanced Micro Devices (AMD) ... 93

Chapter 11 The Silent Contender- IBM Corporation 111

Preface:

There is a big chip war going on today in that part of the semiconductor business that deals with microprocessors. There are three parties to the war. Intel v. AMD has been in the news for some time and their war accelerated in 2003 when AMD took over the x86 (PC style chips) technology lead from Intel with the now famous Opteron family of chips. The other part of the war is between Intel and IBM and this war is much more subtle and quiet but the stakes are also much bigger. IBM doesn't even acknowledge that it is at war and that stance actually is helping the company gain technology advances while operating under Intel's radar scans.

In this book, you will see how and why Intel has enjoyed phenomenal success in the processor chip business. Most notable in Intel's recent history are its Pentium chip models and its Itanium chip project. You will also observe how AMD has been trying to out-Intel Intel for many years and how Intel has consistently been able to swat them down – at least until recently. You will also see how AMD is in a unique position to make the battle with Intel very interesting with its groundbreaking combo 32-bit / 64-bit chip and its leading dual core chip technology.

You will also get a good look at IBM and its rise to the top in powerful commercial processor technology. Additionally you will see how IBM has re-emerged from apparently nowhere to become one of the most powerful chip processor makers of the modern era. Along the way, you will also be introduced to IBM's long-time prowess in creating the most powerful processors that have driven its mainframe computer systems for many years

To cap off the IBM chip success story, you will see how IBM has begun to take advantage of the billions of dollars the company has invested in processor technology over the past fifteen years or so with its RISC processor innovations, 64-bit leadership, and supercomputing initiatives. In the end it will be hard not to notice the rejuvenated IBM with its major agreements with AMD in the x86 industry space, as well as its POWER5 and the coming POWER6 processors in the commercial microprocessor space. Rounding out the battle, you will be introduced to the new phenomenally powerful Cell processor that is being designed for the Sony PlayStation, and you'll see that IBM is the big winner in the next generation game consoles for Sony, as well as Nintendo, and Microsoft.

Though Intel has been the clear leader in chip technology for more than twenty years, there are many reasons why IBM, AMD, and IBM/AMD have a major opportunity to match and/or overtake Intel in the current Intel space, and perhaps even more importantly to steal the opportunity from Intel to become a force in the emerging high tech processor areas such as home control and entertainment consoles, home security, game computers, cell phones, and other high speed embedded processors.

This book about today's processor chip wars is written at a general level for regular human beings, not for computer scientists. Now, I am not suggesting that it would not help to already have a little knowledge and be lightly familiar with some computereeze. Yet the subject matter as presented does not require anything close to IT professional or "Super User" status to understand what the stakes are and who the players are in the 21st century "Chip Wars."

Just a general knowledge of PCs is all you need to get the most out of this book. All along the way, I have tried to humanize the many topics that we cover from the definition of a processor chip to the things that make one chip better than another chip, to the reasons that one semiconductor company may be doing better than another.

This book begins by setting the stage for the chip wars with a historical perspective of the chip itself and a definition of terms, followed by the history of the pioneering companies that formed the semiconductor industry, especially the three companies, Intel, AMD, and IBM that today are battling each other for dominance in the processor chip business. No others need apply.

When you finish this book, you will be in a much better position to know how the chip war games are being played and who has the best opportunity to become the big winner. I wouldn't want to suggest that the knowledge you get from reading this book might affect your future stock transactions, but it has affected mine.

Brian W. Kelly
Wilkes-Barre, Pennsylvania

About the Author

Brian Kelly retired as a 30-year IBM Midrange Systems Engineer in 1999. While with IBM, he was also a Certified Instructor and a Mid-Atlantic Area Designated Specialist. When IBM began to move its sales and support to Business Partners, he formed Kelly Consulting in 1992 as an IT education and consulting firm. Kelly developed numerous AS/400 professional courses over the years that range from soup to nuts.

He has written twenty-five books and numerous magazine articles about current IT topics, including articles for *The Four Hundred, Midrange Computing, Showcase, News/400, AS/400 Systems Management, AS/400 Internet Expert, Computer Business News,* Search400, and others. Kelly has also developed and taught a number of college courses and is currently a member of the Business / IT faculty at Marywood University in Scranton, Pennsylvania, where he also serves as iSeries technical advisor to the IT faculty.

Chapter 1
A Chip Starter Lesson

No Need to Be a Guru

Nobody is going to try to make a computer scientist out of any of the readers of this book who do not already have a technical background. As noted in the preface, this book is not written for the computer technologist but for ordinary human beings who are trying to learn about the essence of the Chip Wars that are in process right now with Intel, AMD, and IBM.

However, just as a book on gardening might rightfully expect some prerequisite knowledge such as the role of soil, sun, and weather; it would not hurt to have a notion of some of the terms used in this book. And, I regret to say there are some things that you may not understand. They are probably the things I do not understand. It's ok. You and I do not have to be engineers to gain the value of this book. But, as you will see, there are some spots where I give an extra explanation for the person who is technical enough to understand it.

This is a book about the business of chip making. This is also about the chip makers. In order to present a true and accurate accounting of what makes whose chip better than whose, there are times that, rather than just beg the argument, I present facts that use words to which most of us are unaccustomed. Even without full knowledge of the technical insights in this book, however, the business underpinnings are the moral of this story.

You should find the treatment of the technical notions in this book to be far more human than scientific. The objective of the section you are now reading is to equip you with the basic tools necessary to understand the premises, the analysis, and the synthesis of the computer chip wars as given throughout this book.

In other words, by learning a few things up-front, you will not have to struggle when such items are used in context. Additionally, if you forget the meaning of any term, you will have a safe place to return to pick up the forgotten information,

Before we present the mini glossary, in a book titled Chip Wars, we need to talk just a little about what a chip actually is.

What is a Chip?

In its most simple form, a chip is a small piece of semiconductor material (usually silicon) on which an integrated circuit (IC) is embedded. A typical chip is less than ¼ square inches in size and can contain millions of electronic components (transistors). Computer systems consist of many chips placed on electronic boards called printed circuit boards.

There are different types of chips. This book is mostly about processor chips. These are also called microprocessors or central processing units – CPUs. Theses chips contain an entire processing unit – in other words, the computer part of the computer, whereas memory chips contain blank memory.

In this book, we examine the various processor chips from the big three vendors, Intel, IBM, and AMD. In order to identify their chip products, these three companies have named their processor chips using some names with which you may already be familiar. All of these chip names are copyrighted but we have spared you the copyright symbols throughout the book to make the book more readable. Here is a sample:

Intel	Pentium, Itanium, Xeon, Celeron, etc.
IBM	PowerPC, POWER5, Cell, etc.
AMD	Athlon, Opteron, Sempron, etc.

More Info on Understanding Chips

Looking at it another way, electronic semiconductors — also known as computer chips, microchips, or integrated circuits (IC) — are the miniature but powerful brains of high-technology equipment. These chips are composed of a myriad of tiny aluminum or copper lines and electric switches, which manipulate the flow of electrical current to achieve specific objectives.

What Is a Semiconductor?

Many non-computerists are able to use the term semiconductor without really knowing what the term actually means. Most of us know that it has something to do with computer chips but that's about it.

Actually, a number of elements are classified as semiconductors. The short list includes germanium, silicon, and zinc. Like aluminum and copper and other metals of which we are aware, these three elements have the ability to conduct electrical current. Additionally, and perhaps more importantly, they can be regulated in the amount of their conductivity.

There are few among us who have not heard of the computer chip or the silicon chip as it is often called. Silicon is the most widely used semiconductor material because it is easily obtained from sand. Silicon (sand) has been used for centuries to make cast iron, bricks, pottery and other common items

Over time semiconductors were used to store and transmit raw data and/or information. Germanium was often the preferred semiconductor used in early designs. The early methods to fabricate circuits however, were quite laborious and inefficient. It was actually the military that demanded more efficient methods of making circuits.

New fabrication methods for integrated circuits were eventually discovered. Thanks to the work of a few Fairchild Gurus (later Intel); Jean Hoerni and Robert Noyce in 1959, the problems of mass producing ICs were solved. Their new process used silicon instead of germanium, and made commercial production of ICs possible. By the end of the 1960s, nearly 90% of all the electronic components manufactured were integrated circuits using silicon as the major semiconductor material.

Computer chips are made and cut from silicon disks of varying sizes, generally eight to twelve inches wide. These disks, called wafers, are thin slices of silicon on which the circuitry of the microchips is layered. Each wafer is eventually cut into many individual chips. See Chapter 3, Chip Fabrication Process to learn more about how chips are made.

The basic process of creating a semiconductor chip wafer uses photolithography, a printing process for creating patterns from photographic images. Using highly automated equipment, workers imprint precise microscopic patterns of the circuitry on the wafers, etch out the patterns with acids, and replace the patterns with metals that conduct electricity. Then, the wafers receive a chemical bath to make them smooth, and the imprint process begins again on a new layer with the next pattern. Wafers usually have from 8 to 20 such layers of microscopic, three-dimensional circuitry. It is really amazing but you don't really need to know much more in order to understand the essence of this book.

Computer chips are produced in semiconductor-fabricating plants. These plants are commonly referred to as "fabs." Intel has a "fab;" IBM has a "fab;" and AMD has a "fab." Actually they have one or more "fabs." Within fabs, the manufacture and cutting of wafers to create semiconductors takes place in "cleanrooms." These are production areas that must be kept free of any airborne matter, because the least bit of dust can damage a semiconductor.

All personnel working in cleanrooms — both operators and technicians — must wear special lightweight outer garments such as Intel's famous "bunny suits." These garments fit over clothing to prevent lint and other particles from contaminating semiconductor-processing worksites. The result of these efforts produces the computer processor chip. For Intel, it may be a Pentium, for IBM, it may be a POWER5, and for AMD, it may be an Opteron.

This is the end of the preliminary information. Unless you have a need to examine the definitions given in the glossary, feel free to move on to Chapter 3.

Chapter 2
Glossary of Chip Terms

When You Need to Understand a Chip Term

Most of the definitions in this section come from other books I have written. However, because there are so many terms in the chip business that do not apply to run of the mill computer technicians and programmers, not to mention the non technical at heart, I added a number of chip terms to the glossary to make it more relevant to the topic at hand. In some cases, I found that the folks on the Web had better wording than I could create so I used some of the wording at Wikipedia at http://en.wikipedia.org or Webopedia at http://www.webopedia.com. Feel free to visit these excellent Web sites for any additional explanations of terms that you may desire. Another site for semiconductor definitions if from BYU at: http://www.ee.byu.edu/cleanroom/EW_glossary.phtml
Of course, the old stand-by, Google will also give back a number of very useful hits on any of these technical terms. Here they are.

Access Time The time it takes for the read/write arm of a disk drive to position itself over the proper record to begin data transfer. Average access time is the time it takes for the read/write arm to move across one-third of a disk, since, on the average, this is the distance the arm must move to find a record.

Access Width The number of bits upon which the Central Processing Unit (processor or CPU) operates during one machine cycle. In an 8 bit machine, the access width is 8; in a 16 bit machine, the access width is 16; in a 64-bit machine, the access width is 64-bits. All other things being equal, systems with a 64-bit access width are 8 times faster than those with an 8-bit access width.

Adapters The circuitry that accommodates the attachment of input / output devices to hardware systems. (See also Interface.)

Address A specific location in computer memory (RAM).

Addressability The largest value that can be stored in the address container in a computer to be able to access computer memory – real or virtual. For example, if we said that the address container were four digits, then intrinsically we would know that the largest address that could be stored would be 9999. So, with a four digit address container, the machine could directly address each memory position from 0 to 9999 (10,000 values) by the number that corresponds in sequence to the memory position. To access memory position 531 for example, the storage container in the computer instruction would hold the address value 531.

Computers of course do not work with decimal digits, they work with binary digits (base 2 instead of base 10), yet computer addressability works the same way. For instance, a computer said to be "32-bit" usually treats memory addresses as 32-bit integers. Thus a computer can address 232 = 4,294,967,296 bytes of memory, or as they say, up to 4 gigabytes. However, older computers often supported memory addresses larger than their word size, or else their memory capacity would be unreasonably small. For instance, the 8-bit MOS Technologies 6502 processor supported 16-bit addresses, or else it would be limited to a mere 256 bytes. Similarly, the 16-bit Intel 8086 supported 20-bit addresses, allowing it to access 1 megabyte rather than 64 kilobytes. A byte-addressable 64-bit computer can address 264 bytes (or 16 exabytes) which as of 2005 is considered practically unlimited. According to Wikipedia, this is far more than the total amount of RAM ever manufactured.

Alphanumeric Character An alphabetic, numeric, or special character such as (,.<>?/$%&) – also called alphameric.

ALU Arithmetic /Logic Unit. That area of a CPU that provides mathematical and logical capability. An arithmetic/logic unit (ALU) is a core component of all computer CPUs. ALUs are capable of calculating the results of a wide variety of basic arithmetical computations.

Analog Computer A computer that measures sensory input and provides sensory output in the form of different strength electrical signals. Opposite of digital computer which uses discrete voltages as 1's and 0's to determine the absence or presence of data. Most often used for process control applications such as temperature monitoring.

Analog to Digital Converter A separate box or an adapter on a data processing device that converts analog signals to digital signals.

ASCII American Standard Code for Information Interchange. An 8-bit internal coding structure, which has been adopted as standard by ANSI. The coding structure most commonly found in Unix and personal computer systems.

ASIC Application Specific Integrated Circuits. ASIC chips are the heart of the technology used in wireless phones and infrastructure as well as direct broadcast satellite (DBS) systems, very small aperture terminals (VSATs), radar systems, digital and mobile radios, synthesizers, voice storage systems, security systems and instrumentation.
ASICs are chips designed by customers for specific applications by integrating cells from a standard library of pre-tested code. ASIC design is faster than designing a chip from scratch, and design changes can be made more easily.

BIOS Basic Input Output System. Like an input/output control system. A set of programs written to support the attachment of specific devices to specific computing systems.

Bit **A** single binary signal -- a 0 or a 1 -- is called a "bit" (BInary digiT). A bit has only two states, 0 or 1 (OFF or ON), so one bit can only represent two possible values. But you can use multiple bits together to represent larger numbers, just like you can use multiple decimal digits to represent numbers greater than 9, (e.g. three digits to represent values up to 999). You compose large binary values using the same concepts you use to compose large decimal (base 10) values, except you are working in base 2 notation, not base 10.

Computers operate using "binary circuitry" for everything they do. Binary circuits have only two states (or conditions), OFF or ON. So everything a computer does to be represented by some combination of OFF and ON signals. The OFF state is denoted by the value 0 and the ON state by the value 1. Because there are only two possible states, there are only two values in the binary number system, 0 and 1. So all data stored in a computer is represented by combinations of 0s and 1s.

Boot The process of starting a computer system and having its bootstrap load into memory. On larger systems it is known as Initial Machine Load - IML, or Initial Program Load - IPL.

Branch Computer systems differ from calculators in that they hold many instructions and some of the instructions are logical instead of arithmetic. For example, the computer may compare to see if the data being used pertains to a female or a male and take different paths or branches to the section of the program that deals with the male options or the female. Technically, a branch is a point in the instruction stream of a computer program where the address of the next instruction is not the next sequential storage location. A branch may be unconditional (implying that the branch is always taken) or conditional, implying that the decision to take the branch or not depends on some condition that must be evaluated, such as the resolution of the data being for a male or female.

Branch predictor Part of a processor that examines the instruction flow to determine whether a conditional branch of a program is likely to be taken or not. This process is called branch prediction. Branch predictors are crucial in today's modern, superscalar processors for achieving high performance. They allow processors to fetch and execute instructions without waiting for a branch to be resolved.

Bulk CMOS A term used to differentiate old style CMOS implementations from the newer silicon on insulator (SOI) style. The circuitry is implemented on a standard bulk Silicon wafer rather than in a thin layer of active Silicon on insulator (SOI substrate). See CMOS

Bus An electronic means of connecting large numbers of devices together in a computer system. Typically involves sockets which permit easy-to-use connections to solid logic cards. It provides a common access capability for computer hardware.

Byte A group of 8-bits plus a check bit. Provides for 256 different combinations of letters and numbers and it is the measure of storage for alphanumeric character of information in memory and in disk.

Chip A small square piece of silicon which contains miniaturized electronic circuitry. Size of an average chip is 1/8" square. A commonly used term for integrated circuit (IC).

CISC Complex Instruction Set Computing. The traditional computer architecture that includes a rich set of complex instructions such that one instruction can perform many tasks. Used in mainframes and Intel Pentiums processors.

Client A computer or software program that accesses resources over the Internet. There are thin clients (browsers) and fat clients (client server applications such as email) It is also an application that requests a server to perform a function, and cooperates with the server application.

Clock That part of the CPU which provides the timing pulses to synchronize system activity. Each time the clock "pulses" the computer is capable of performing one task. The terms megahertz (MHz or millions of machine cycles per second) and gigahertz (GHz or billions of machine cycles per second) is a measure of CPU clock speed.

CMOS Complimentary Metal Oxide. The conventional technology for chip building that existed prior to Silicon on insulator (SOI).

Coding See programming

Component A dependent functional part of a hardware or software system. For example, memory is a component of a CPU; data management is a component of an operating system.

Computer An electronic machine which stores its own instructions and is capable of the following: accepting input through attached devices; processing in the form of addition, subtraction, multiplication, division, and comparisons. It presents the results of processing in some output format.

Computer architecture The theory behind the design of a computer. In the same way as a building architect sets the principles and goals of a building project as the basis for the draftsman's plans, so too, a computer architect sets out the computer architecture as a basis for the actual design specifications.

Control Section of CPU That part of a CPU which decodes and executes one instruction at a time, sends the proper signals to the other parts of the CPU (registers, ALU, memory) to carry out required operations, and governs the flow of all operations within the CPU.

Core Storage An outdated memory technology consisting of nonvolatile ferric oxide coated magnetic cores strung on a matrix of thin wires.

Core The execution unit in a CPU. For example, in dual core systems, there are two processors or execution units.

CPU The Central Processing Unit of a computer system. Consists of a control section, an arithmetic/logic unit (ALU), memory, and high speed registers.

Cycle An interval of time (which varies based on CPU clock speed) during which, a quantifiable number of CPU (processor) activities can occur.

Data Representation The internal binary coding structure of a data processing machine.

Double data rate (DDR) SDRAM **A** later development of SDRAM. This type of memory has been used as PC memory from 2000 onwards

Depleted CMOS A variation of the SOI technique that uses an ultra-thin active silicon layer. The full explanation of this is usable only for a semiconductor engineer. In 2005, IBM used and AMD used a partially depleted CMOS notion while Intel, who were proponents of bulk SMOS vs. SOI in the first place were rallying around the notion of fully depleted CMOS. The battle continues.

Digital Computer Uses constant strength electrical current to represent "on" conditions, and no current to represent "off" conditions. On conditions translate into binary 1's, and off conditions translate into binary 0's. The determination of "on" and "off" is based upon the presence or absence of a pulse within a time interval.

Digital circuits are also those associated with two states, on and off. On is represented by some constant level of voltage and off is no voltage or a different but constant value. In essence, they are electric circuits based on a number of discrete voltage levels. In most cases there are two voltage levels: one near to zero volts and one at a higher level depending on the supply voltage in use. From the absence or presence of voltage, values are discerned.

Digital signal processor (DSP) A specialized microprocessor designed specifically for digital signal processing, generally in real-time. Though DSPs can also be used to perform general-purpose computation, they are not optimized for this function. DSP function can also be implemented via pure software. A typical use for a DSP would be to re-encode audio for Internet radio stations in real-time.

Direct Rambus DRAM (DRDRAM) Often called RDRAM, DRDRAM is internally similar to DDR SDRAM, but uses a special method of signaling developed by the Rambus Company that allows faster clock speeds. RDRAM chips are packaged on modules called RIMMs, which are not compatible with the DIMMs used for plain SDRAM. Intel licensed the Rambus technology and the company has introduced chipsets with RDRAM support. Early P4 systems could only use RDRAM, but as prices remained high, Intel finally introduced support for DDR. Sony used RDRAM in its PlayStation 2 video game console, and announced it would use Rambus's XDR memory in its PlayStation 3, expected in 2006.

DRAM A type of random access memory that stores each bit of data in a separate capacitor. The number of electrons stored in the capacitor determines whether the bit is considered 1 or 0. As the capacitor leaks electrons, the information gets lost eventually, unless the charge is refreshed periodically. Because it must be refreshed periodically, it is a dynamic memory as opposed to SRAM and other static memory. Also, since DRAM loses its data when the power supply is removed, it is in the class of volatile memory devices. DRAM is also in the class of solid-state memory.

Dual core A dual-core CPU or processor chip combines two independent processors (two computers) and their respective caches and cache controllers onto a single silicon die, or integrated circuit – i.e. on one chip.

EBCDIC Extended Binary Coded Decimal Interchange Code. Developed by IBM, an internal coding structure which defines a character for each of the 256-bit configurations available in an 8-bit byte.

Emulation The imitation of one computer by another, or one chip by another primarily by using hardware.

EPROM Erasable Programmable read only memory. A type of PROM which can be erased and reprogrammed.

Execute To perform, or carry out an instruction or series of instructions; Run.

Execution Unit In computer engineering, an execution unit is a part of a CPU that performs the operations and calculations called for by the program. It often has its own control unit, registers, and other electronics, such as an arithmetic and logic unit or floating point unit, or some smaller, more specific components. It is commonplace for modern CPUs to have multiple parallel execution units, referred to as scalar or superscalar design.

General Purpose Computer A computer capable of performing a variety of different functions. Each program gives a general purpose computer a different set of instructions, and therefore a different purpose.

Gigabyte (1 Billion bytes)

Gigahertz One billion hertz (machine cycles per second)

Hardware The physical components and peripheral equipment which are part of a computer system.

Hyper-Threading An Intel developed technology that permits multiple threads or processes to run on one processor. Thus, for the dual-core Intel chips that deploy Hyper-Threading, four processes can be going on within one chip simultaneously. When the OS and the applications support the splitting of work into multiple concurrent threads, then substantial performance improvements can be achieved in performance. This technology does require a computer system (PC or Server) with an Intel Pentium processor supporting Hyper-Threading Technology and an HT Technology enabled chipset, BIOS and operating system.

Performance will vary depending on the specific hardware and software you use. For more information on Intel Hyper-Threading, feel free to visit their website at http://www.intel.com/info/hyperthreading/ for more information including details on which Intel processors support HT Technology. See also multithreading

I/O Input/output devices. The process of input and output in a computer system. That part of computer processing that deals with devices that are not on the chip, such as printers, disk drives, terminals, and networking.

I/O Interface The electronic circuitry which provides the necessary function to allow data processing devices to be connected.

I/O Ports Input/output ports. The physical connection points on a computer system for the attachment of Input/Output devices.

IC Integrated Circuit. Also referred to as a chip.

IML Initial Machine Load. a.k.a. Initial Program Load or IPL. The process of starting a computer system and having its bootstrap program load. On smaller systems it is known as Boot or Re-Boot.

IPL Initial Program Load. a.k.a. Initial Machine Load or IML. The process of starting a computer system and having its bootstrap program load. On smaller systems it is known as BOOT or Re-Boot.

Interface See I/O Interface.

Interrupt On a computer system, a capability which allows a program, component, or peripheral device to interrupt the processor to notify it that some action has been completed or needs to be performed.

IOCS Input/Output Control System. A set of programs written to support the attachment of specific devices to specific computing systems.

Instruction A way of telling the computer to do one specific thing. A specific request for processing action in a machine.

Instruction set Sometimes referred to as instruction set architecture (ISA). The instruction set of a computer in many ways differentiates it from all other computer types. It describes the aspects of a computer architecture visible to a programmer, including the native data types, instructions, registers, addressing modes, memory architecture, interrupt and exception handling, and external I/O (if any).

An ISA is a specification of the set of all binary codes (instruction operation codes or op-codes) that are the native form of commands implemented by a particular CPU design. The set of op-codes for a particular ISA is also known as the machine language for the ISA. For standard Intel processing for example, there is the x86 instruction set, which is property of Intel and is licensed by AMD so that AMD can use the Intel designed instruction set. The IBM Power Architecture instruction set is designed by IBM and owned by IBM but usable to the POWER.ORG partners.

iSeries IBM Business Server - follow-on to AS/400 and predecessor to i5. The iSeries is built using IBM's POWER processors.

K A symbol used to represent 1024 storage positions. Computer systems operate m binary, or powers of 2, 2^{10} = (1024)- the closest power of 2 to 1000, or k; an abbreviation for kilo, which means 1000.

Load The process of bringing a program into computer memory prior to execution.

Location A specific address within computer memory where a piece of data is stored.

Low-k dielectric Material used to insulate adjacent metal lines (interlayer dielectric, ILD) in advanced integrated circuits. A low k factor reduces undesired capacitive coupling, and hence "cross talk", between lines.

Machine Cycle See Cycle.

Machine Language The native language of a computer system, used directly by the machine – such as the x86 instruction set..

Main Storage That memory portion of the CPU which stores instructions and data during program execution.

Mainframe A large computer system usually costing in excess of $100,000.

Megabyte One million bytes

Megahertz One million hertz (machine cycles per second)

Memory See Main Storage, RAM.

Microcomputer A small processor which serves as a satellite processor in larger systems and functions as the heart (CPU) of a personal computer system.
Microprocessor See Microcomputer.

Microsecond One millionth of a second.

Millisecond One thousandth of a second.

Minicomputer A small computer system from the 1970's and 1980's, usually costing between $10,000 and $250,000.

MirrorBit A proprietary AMD architecture that enables a Flash memory product to hold twice as much data as standard Flash memory, without compromising overall performance.

Monolithic Storage Storage made up of monolithic integrated circuits.

Moore's Law The observation first made by Intel co-founder, Gordon Moore that semiconductor density doubles roughly every 12 to 18 months.

MTBF Mean Time Between Failures. The average time between failures on a data processing device.

Multiprogramming An operating system software technique which provides an environment for the execution of more than one program at a time in the same computing system. In such programming systems, since there is only one thread, the processor is shared by all tasks. Tasks do not run simultaneously.

Multiprocessing A chip capability that enables two simultaneous processes to run on the same chip in two processors. It refers to the utilization of multiple CPUs in a single computer system or on a single chip. . This is also called parallel processing.

Multicore chips Multicore chips are the biggest change in the PC programming model since Intel introduced the 32-bit 386 architecture," When using multicore chips (dual core chips are multicore chips with just two cores, the operating system most often controls the overall assignment of tasks to the various threads and the cores. Either the OS or a multithreaded application parcels out work to the multiple cores. When a multicore processor has completed a task, one core takes the completed data from the other cores and assembles the final result.

A multicore architecture is actually a SMP (symmetric multiprocessor) implemented on a single VLSI (very large scale integration) circuit. Multiple CPUs enable true parallelism. The goal is to allow greater utilization of thread-level parallelism (TLP), especially for applications that lack sufficient instruction-level parallelism (ILP) to make good use of superscalar (modern powerful processor chips) processors.

When one chip handles the entire load it is called Chip-level multiprocessing (also known as CMP). To effectively use multicore (including dual core versions), vendors must redesign their programs so that the processor can run them as multiple threads. It is substantially more challenging to create software that is multithreaded.

To be effective in optimizing applications, programmers must find good places to break up the applications, divide the work into roughly equal pieces that can run at the same time, and determine the best times for the threads to communicate with one another.

IBM released the industry's first dual-core server chip, the Power 4, in 2001. Last year, In 2004, IBM introduced the dual core POWER5, which runs four times faster than its predecessor. In 2005, AMD and Intel joined the foray with their x86 dual core offerings.

Multimedia The use of several different media to convey information (text, audio, graphics, animation, video, and interactivity). Multimedia also refers to computer media. As the information is presented in various formats, multimedia enhances user experience and makes it easier and faster to grasp information. Presenting information in various formats is nothing new, but multimedia generally implies presenting information in various digital formats using a GUI with video and audio fro effect.

Multithreading A property of a computer system or computer chip that enables multiple portions of programs to execute concurrently. Multiple threads typically share the state information of a single process, and share memory and other computer resources directly. This is in contrast t the typical notion of multitasking operating system processes in that such processes are typically independent, carry considerable state information, have separate address spaces, and interact only through special system-provided inter-process communication mechanisms. Multithreading is a typically a better performing option than multitasking, especially in single chip machines since all the action occurs on the chip.

Nanosecond One billionth of a second.

Network A data communications term used to represent a configuration in which two or more terminal or CPU locations are connected.

Node One drop on a network; one location; one computer or terminal or other network device.

Non-Volatile A characteristic of certain memory technologies which provides for the contents of memory to be retained when power is turned off- i.e., power is not required to maintain the data contents.

Normalization A relational design technique used to help better database record designs.

OEM Original Equipment Manufacturer.

Operating System (OS) Control software which governs all activities in a computer system, such as scheduling, resource allocation, data management, I/O control, etc. Operating systems are the key ingredient in personal computer systems. They make a computing system easy to use (friendly) or not easy to use (unfriendly). Examples are Windows, i5/OS, OS/400, Unix

Out of order execution Out-of-order execution is a technique used in most high-speed microprocessors in order to make use of cycles that would otherwise be wasted by a certain type of costly delay. Almost all modern CPU designs include support for out of order execution.

Output Device A machine capable of receiving information from a CPU, and based on the type of output device, producing hard copy, display, magnetic, or other forms of output.

Parallelism; The notion in a computer system or a computer chip to execute multiple programs or parts of the same program in parallel (at the same time). There are different aspects to parallelism. Thos with parallelism in their title are defined below

> **Instruction-level parallelism** (ILP) is a measure of how many operations in a computer program can be performed simultaneously.
>
> **Thread-level parallelism** (TLP) is the notion of parallelism implemented at the application level. It is possible when applications are written to run multiple threads at once. Commercial databases often use multiple threads to achieve high performance. By running many threads at once, applications are able to tolerate the high amounts of I/O and memory system latency their workloads may encounter.

pSeries IBM RISC/Unix based server - RS/6000 follow-on

PSRAM A dynamic RAM with built-in refresh and address-control circuitry to make it behave similarly to static RAM (SRAM). It combines the high density of DRAM with the ease of use of true SRAM.

Peripheral Equipment Devices that attach to systems but are not integral to them. They provide the system with additional facility- e.g., a printer, a display, a hard disk drive, a voice synthesizer. Commonly referred to as "peripherals."

Performance Together with "ease of use," one of the two major factors upon which the total productivity of a system depends. Performance is largely determined by a combination of three factors: throughput, response time, and availability. With CPU chips, standard benchmarks are conducted to determine relative speed of one chip to another.

Personal Computer A microcomputer designed for ease of use and packaged with user friendly software.

Pipelining (Instruction Pipelining) Instruction pipelining uses a form of instruction level parallelism for increasing the throughput of a digital circuit, particularly a CPU. The basic idea is to divide the logic into stages, and to have the CPU work on different data within each stage. Wikipedia (http://en.wikipedia.org) uses a laundry analogy to explain the notion of pipelining. "An often used real-world analogy involves doing the laundry: if you have two loads of laundry to do, you can either wash the first load and then dry the first load, before moving onto the next, or, you can wash the first load, and when you put it in to dry, you can put the next load in to wash. If each step takes 20 minutes, then you will finish in 60 minutes instead of 80." When a chip designer pipelines a digital circuit, it works the same way. Data enters the first stage, and takes some time to process. When the data finishes the first stage, the clock ticks, and the intermediate results are latched into registers at the head of the next stage, while the next set of data enters the beginning of the first stage. Ideally, pipelining increases throughput by a factor equal to the number of stages used. Realistically, the time taken by the extra logic added (in the form of latches or registers) to store the intermediate values results in diminishing returns, but still typically lots better than not using pipelining.

Power Supply An integral part of a personal computer system, the part which provides the required voltage and current.

Primary Storage See Main Storage.

Priority A numeric value representing the relative execution importance of a job. Lower numbers mean higher priority. When a task with the lowest numbered (highest) priority wants the processor, it is preemptively dispatched to that task - even if the processor is doing lower priority work at the time. Priority provides the divvying rules for sharing the central processor resource, when jobs are competing for CPU time.

Processing The action of CPU operations on data for the eventual production of output.

Program A set of stored instructions which tells a computer system what operations are to be performed and in what sequence to perform the set of instructions.

PROM Programmable Read Only Memory-a non-volatile memory technology which can be programmed after manufacturing, by the end user, but cannot be erased and reprogrammed.

Programming The process of understanding program requirements, coding source instructions in a computer language, compiling source into machine language, testing and debugging with representative data, verifying the results, and repeating the process until correct.

RAM Random Access Memory. The most common type of volatile main storage used in small computer systems.

Real time The processing and recording of data transactions as they occur.

Register A processor register is a small amount of very fast computer memory used to speed the execution of computer programs by providing quick access to commonly used values—typically, the values being in the midst of a calculation at a given point in time. These registers are the top of the memory hierarchy, and are the fastest way for the system to manipulate data. Registers are normally measured by the number of bits they can hold, for example, an "8-bit register" or a "32-bit register" or a "64-bit register."

Register renaming A technique used to avoid unnecessary serialization of program operations imposed by the reuse of registers by those operations. For the neophyte is helps to know that it is a technique used by programmers so that they can write code that executes in parallel.

Resolution In computer graphics, a measure of the clarity or sharpness of a produced image typically expressed in the number of points (pitches) in a horizontal and vertical matrix.

RISC Reduced Instruction Set Computing. A computer architecture developed by IBM's John Cocke that reduces the number of instructions on a processor chip so that each remaining instruction can be made to operate at a substantially faster speed.

ROM Read Only Memory. A non-volatile memory technology which is programmed during the manufacturing process and cannot be changed by the end user.

SAP The name of one of the largest ERP software vendors. SAP is also used as a nickname for the company's R/3 package.

Synchronous Dynamic RAM (SDRAM) An improved type of DRAM. DRAM has an asynchronous interface, meaning that it reacts immediately to changes in its control inputs. SDRAM has a synchronous interface, meaning that it waits for a clock pulse before responding to its control inputs. The clock is used to drive an internal finite state machine that can pipeline incoming commands. This allows the chip to have a more complex pattern of operation than plain DRAM.

Security Prevention of access to or use of data or programs without authorization.

Semiconductor A term used to refer to electronic circuitry contained on a silicon chip.

Server. An entity that provides a network service. A server can be hardware (such as a file server), software (such as a mail server), or services (such as a transportation service). A mail server is a program that accepts, relays, and delivers mail.

Silicon-On-Insulator (SOI) It is a process and a physical item. It is basically a silicon wafer with a thin layer of oxide ($SiO2$) buried in it. Transistors and various other on-chip devices are built into a layer of silicon on top of the buried oxide. The SOI substrates provide superior isolation between adjacent devices in an integrated circuit as compared to devices built into bulk wafers. The process also provides improved chip performance of SOI devices due to reduced parasitic capacitances. In other words, SOI is good for chip making. Those who do not deploy it, do so at their own peril.

Simulation The use of software programming techniques to imitate (duplicate) the operation of one computing system on another computing system. Contrast with emulation.

Simultaneous Multithreading (SMT) permits multiple independent threads to operate simultaneously rather than on a time slice basis. Additional hardware is required on the chip to accomplish this. This technique reduces the impact of a thread stalling for several machie cycles when it must access another resource such as memory.

Software A set of stored programmed instructions that tells a computer system what to do. Contrast with hardware.

Software Developer A person or company that develops and markets software.

Solid Logic Technology (SLT) A technological advance which miniaturized and refined the components of the "second generation" of computers and brought about the "third generation."

Speculative Execution A technique implemented by chip vendors trying to achieve better parallel performance with multiple threads and/or multiple CPUs on a chip. It is the execution code in a parallel thread whose result may not actually be needed at all. It is an optimization to attempt to achieve the best performance from parallel threads. It is useful only when early execution consumes less time and space than later execution would, and the savings are enough to compensate, in the long run, for the possible wasted effort of computing a value which is never used. It is one of a number of techniques that IBM has implemented to get top performance from its chips in lieu of using high clock speeds.

Speed A performance measurement of a CPU, typically specified in terms of "cycle time" or Megahertz (MHz).

SRAM Short for static random access memory. SRAM is a type of memory that is faster and more reliable than the more common DRAM (dynamic RAM). The term static is derived from the fact that it doesn't need to be refreshed like dynamic RAM. While DRAM supports access times of about 60 nanoseconds, SRAM can give access times as low as 10 nanoseconds. In addition, its cycle time is much shorter than that of DRAM because it does not need to pause between accesses.

Unfortunately, it is also much more expensive to produce than DRAM. Due to its high cost, SRAM is often used only as a memory cache.

Strained Silicon The term strained silicon refers to chip transistors built with strained silicon. The silicon is "stretched out" because of the natural tendency for atoms inside compounds to align with one another. When silicon is deposited on top of a substrate with atoms spaced farther apart, the atoms in silicon stretch to line up with the atoms beneath, stretching -- or "straining" -- the silicon. In the strained silicon, electrons experience less resistance and flow up to 70 percent faster, which can lead to chips that are up to 35 percent faster -- without having to shrink the size of transistors.

Strained Silicon with DSL In December, 2004, IBM and AMD's cooperative agreement bore fruit in terms of an enhancement to strained silicon. The improvement known as "Dual stress liner," helped ensure that Moore's Law applies to more than just the chips that run supercomputers. The notion if DSL is that fewer electrons leak away in transit when transistors are turned on

Storage Term used for memory on the system as well as disk. Main storage is memory and auxiliary storage is disk. All programs execute in memory. On PCs, this is often called RAM.

Superscalar a CPU architecture implemented with a form of parallelism on a single chip, thereby allowing the system as a whole to run much faster than it would otherwise be able to at a given clock speed.

The simplest processors are scalar processors. A scalar processor processes one data item at a time. In a vector processor, by contrast, a single instruction operates simultaneously on multiple data items. The difference is analogous to the difference between scalar and vector arithmetic. A superscalar processor is sort of a mixture of the two. Each instruction processes one data item, but there are multiple processing units so that multiple instructions can be processing separate data items at the same time.

Super-threading A form of simultaneous multithreading, similar in design to hyper-threading. In super-threading, the processor can execute instructions from a different thread each cycle. Thus cycles left unused by a thread can be used by another that is ready to run. Still, a given thread is almost surely not utilizing all the multiple execution units of a modern processor at the same time. More advanced implementations of SMT allow multiple threads to run in the same cycle, using different execution units of a superscalar processor.

Symmetric Multiprocessing (SMP) A multiprocessor computer architecture where two or more identical processors use a single shared main memory. Most common multiprocessor systems / chips today use an SMP architecture.

Synchronous graphics RAM (SGRAM) A specialized form of SDRAM for graphics adaptors.

System A group of related parts or components working together as a whole for the accomplishment of an objective. Computer system.

System Software The set of operating system programs, utility programs, and languages, which gives a computer system its basic "personality.

Terminal A device, usually equipped with a keyboard and display, capable of sending and receiving information over a communications channel.

Testing The process of executing programs using test data with predetermined results.

Thread The basic unit to which a processor allocates execution time. A part of a program that can run independently of other parts. Operating systems and chips that support multithreading enable programmers to design programs whose threaded parts can execute concurrently.

Throughput A measurement of the actual work a computer system can perform.

Time Sharing A method of using a computing system that allows a number of users to execute programs concurrently and to interact with the programs during execution. A personal computer system when equipped with a communications adapter can be a terminal in a time sharing network.

Time slicing In systems with time sharing, the allocation of time slices to user jobs.

Voice Synthesizer A peripheral device which can synthesize voice from stored data, as instructed by a computing system.

Virtual Storage A concept which divides memory into pages and uses disk as primary storage for pages, thereby allowing programs much larger than available main memory to execute.

VLSI Very-large-scale integration (VLSI) of systems of transistor-based circuits into integrated circuits on a single chip first occurred in the 1980s as part of the semiconductor and communication technologies that were being developed Today all this happens on a single chip.

Volatile Memory Main storage which requires continuous electrical power to retain its contents. When power is turned off, the contents of volatile memory are destroyed.

User Anyone who makes use of the services of a computing system.

VRAM is a dual-ported version of DRAM formerly used in graphics adaptors. It is now almost obsolete, having been superseded by SDRAM and SGRAM.

Word A set length of data, in terms of the number of bits, which a computing system acts upon in normal operations. Popular word lengths are 8-bits, 16-bits, 32-bits and 64-bits. Occupies one location in the computer's memory. Word is also a nickname for Microsoft's word processing software.

xSeries IBM Intel based PC server

zSeries IBM Mainframe based server

Chapter 3
The Chip Making Process

Much More Than a Soap Opera

"Like sands through the hourglass, so are the days of our lives." Before and after Macdonald Carey was the announcer for "Days," I can recall hearing the "Soap" in the background as I tended to other duties in my childhood home and now in my own home. I guess the saying means that there is nothing more plentiful than sand and there are also many days in our lives – and perhaps that each grain of sand and each day of our lives are different.

There is not one invention that has affected the days of our lives as much as the integrated circuit that is indirectly from sand. In fact, the silicon made from sand is the basis for the semiconductor industry of today and it serves as the basic ingredient upon which processor chips are built.

Let's take a brief time-out now to discuss the generic process of taking sand and creating integrated circuits in silicon chips. After all, sand is the basis upon which the multi-billion dollar semiconductor business was built and there would never be a "Chip War" going on today if it weren't for sand – no matter what MacDonald Carey may or may not have said of the process.

Sand to Ingot to Wafer

The creation of silicon wafers is a science and technology of its own, independent of the memory and processor chips that ultimately are fabricated. The creation of the first wafers goes back to the early 1960's and perhaps even the late 1950's.

It was in the mid 1970's as I recall when IBM, a leader in chip technology at all levels at the time, and one of the few companies that made its chips with 8 inch (200mm) wafers, chose to outsource its wafer building. It had become increasingly expensive to manufacture wafers when the needs of the business could not use all the wafers that were produced. Since that time IBM, along with most chip makers have purchased their wafers from a number of outside sources.

The wafer making companies build their 8 inch (200mm) or 12 inch (300mm) paper thin wafers, also called substrates, by using sand and some other materials to first "grow" large silicon ingots which are then sliced like lunchmeat and eventually sold to chip manufacturers

The semiconductor wafer manufacturing process begins then with one of the most common elements on earth, silicon. Silicon of course is found in abundance in sand, but before it is used in semiconductor manufacturing it is refined to be virtually 100% pure. Purity of materials is fundamental to delivering chips that function as intended.

Wafer manufacturers go through great pains to assure this purity in order for their customers to be able to build chips with no issues.

Growing a silicon ingot can take anywhere from one week to one month, depending on many factors, including size, quality and the end users specifications. There are a number of different technologies deployed to build the ingots. In most processes, as a general rule, chunks of silicon are combined with small quantities of elements called dopants in order for the silicon to achieve the right electrical properties in the grown ingot.

Figure 3-1 Six inch Silicon Ingot Courtesy of TI

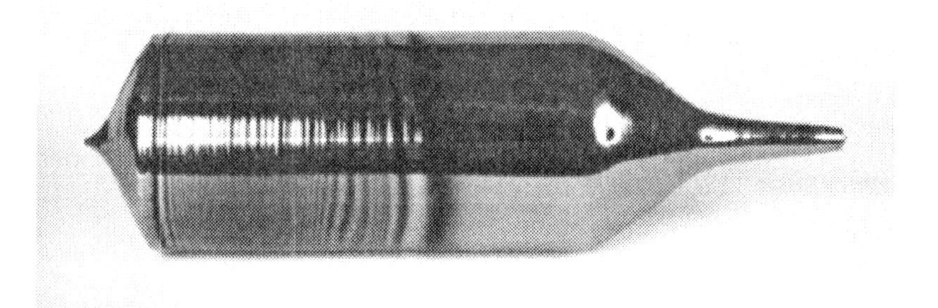

The kluge of materials is then heated to a temperature above the melting point of silicon, 1420 degrees Celsius. Once conditions for the crystal growth have been met, the ingot begins to grow until the diameter of the "crystal" increases to the desired diameter. Once the ingot is fully-grown, it is ground to a rough size diameter a little larger than the desired diameter of the finished silicon wafer. Once it has passed a number of inspections, the ingot is sliced into wafers by a diamond edge saw just a little thicker than the desired specification.

The accuracy of the diamond saw also helps to minimize damage to the wafers such as thickness variations and bow and warp defects. Other processes continue on the wafer to remove saw marks and surface defects from the front and backside of the wafer and to thin the wafer. There are also processes that help to relieve any molecular stress accumulated in the wafer from the slicing process.

Figure 3-2 Sliced Ingot & Wafers

Following the "smoothing process," the wafers are etched and cleaned using various chemicals including some nitric acids. This helps alleviate any microscopic cracks and other surface damage that might have been caused by the smoothing process.

As a next critical step, the wafer must go through a process called edge grinding to round the edges. The edge grinding tests the durability of the wafer among other things and the process drastically reduces the probability of further breakage. When the edges have been rounded, they are often polished, which again improves the overall cleanliness and it actually reduces the probability of breakage by another 400%.

Though it seems like we've already done enough to this poor little wafer, there is one final and most crucial step that must be done. The wafer must be polished. For this, the wafer is transferred into a very clean room.

> Note: Clean rooms are rated and range from Class 1 to Class 10,000. The number corresponds to the number of particles per cubic foot. For reference, these particles are not visible to the naked eye and in an uncontrolled atmosphere, such as a living room or office, the particle count would likely be 5 million per cubic foot. To help maintain this level of cleanliness, the workers must wear clean room suits that cover their body from head to toe and are designed to not collect or carry any particles. They also will stand in a vacuum that blows away any small particles that might have accumulated before entering the room.

Figure 3-3 Clean Room Cutaway

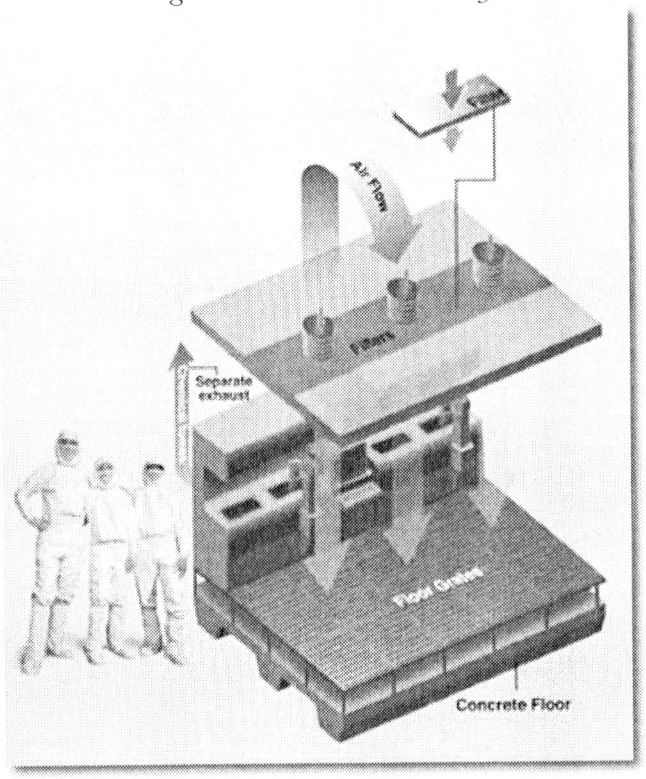

Most often, there are two to three different stages of polishing using progressively finer slurry or polishing compound. The majority of the times, wafers are polished on the front side only, excluding 300 mm wafers, which are double side polished. When polishing is complete, the sides that have been treated are so clean and smooth, they look like mirrors. For normal "bulk silicon" chip making, the "mirror side" of the wafer is used for device fabrication (to build circuits). Therefore, this surface must be free of any issues such as micro-cracks, scratches, or residual work damage. After polishing, there is a final cleaning to remove even the smallest particles.

As what might seem to be a special treat for the wafers after all they have been through, these clean silicon disks are then finished in clean baths. Following the bath, they are sorted by vendor specifications, and inspected under high intensity lights or laser-scanning systems.

This helps further detect unwanted particles or other defects. Once they have passed this series of inspections, the finished wafers are packaged in a holder, sealed with tape, and then placed in a vacuum-sealed plastic bag, and then an air tight foil outer bag. All of this protection is to assure that no particles or moisture enters the holder upon leaving the clean room.

The Next Step for the Wafers

Chip manufacturers such as IBM, AMD, Texas Instruments, and others buy their wafers from companies that specialize in manufacturing the silicon ingots and slicing them appropriately. Once the bare wafers, or substrates as they are called, arrive at the receiving dock the chip manufacturer's real work begins!

Silicon is used as the base material for chips precisely because it is cheap and it conducts electricity at the right level for its use in computer chips. Under certain conditions silicon conducts electricity and under other conditions it does not. That's why it is listed under the term "semiconductor." This on/off capability is what underlies the transistor switching action that forms the ones and zeros of digital logic.

Just as there are a lot of steps in creating the wafer itself, there are even more steps in semiconductor manufacturing in order to build all of the tiny components with the proper electrical structure to rapidly switch and transfer signals. In addition to the switching transistors and the metal traces (copper or aluminum) that conduct electrical signals between various regions of the chips, insulating materials separate conducting areas of the device in much the same way as electrical wire is protected by a non conductive coating. .

In order to alter the characteristics of the semiconductor to provide the circuits necessary, a number of steps are undertaken in the manufacturing process in various sequences, depending on the complexity and functionality of the device. The technical terms for a number of these sequences are as follows:

- ✓ **Deposition**
- ✓ **Diffusion**
- ✓ **Ion Implantation.**

None of us are going to be Physics professors after this brief chip building lesson, but we should all have a better notion of what goes on in the chip building process. By being introduced to a subset of the many steps to build a "simple chip," you will be in a much better position to understand why it takes so long to go from design to final shipment. Let's now define the three terms introduced above:

- **Deposition** is the process by which an insulating layer is grown on the silicon substrate
- **Diffusion** bakes impurities into areas of the wafer to alter its electrical characteristics
- **Ion implantation** is another process for infusing the silicon with various chemicals called dopants to change its electrical characteristics

As they say so often in the business world, to net it out, the process involves putting stuff on the chip, baking that and other stuff into the chip to help its electrical characteristics, and mixing more silicon and chemicals on the chip to form a layer.

In between these steps, areas of the chip are patterned with an image for that particular layer of the device using a process called photolithography. In many ways the process of photolithography is the same process used at a print shop in that a picture of an image is reproduced. The circuitry lines on a chip are so small that a very precise "mask" is used to expose photoresist, a very light sensitive material that has been applied across the wafer, very much like emulsion on film. This pattern hardens into an exact representation of the mask when it is developed.

The next process called etching then removes selective areas of the pattern using plasma that reacts to the material that is not covered by the "single surface chip picture" that was built by the hardening of the photoresist substance.

Figure 3-4 Photolithography

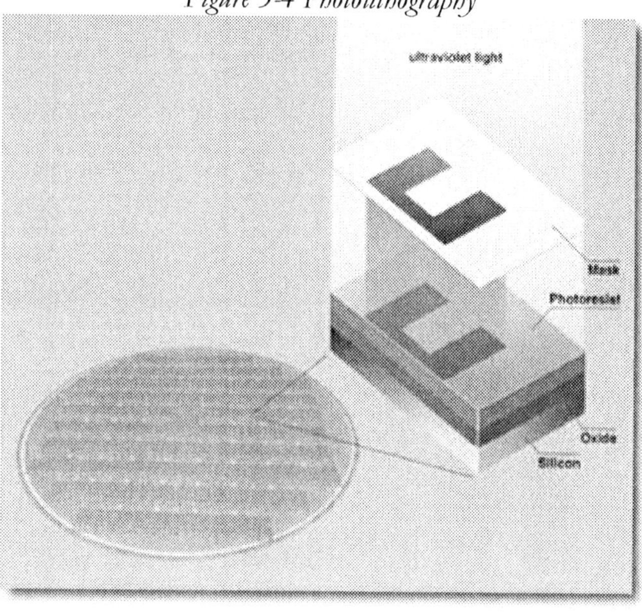

If a wafer / chip had just one layer, the job might be over now, other than for the cleaning but chips contain many miniscule layers. So, this process gets repeated and repeated and repeated to create layers of transistors with precise operating characteristics. The characteristics of course get determined by the deposition, diffusion and ion implantation steps for each chip layer. A specialized deposition process called "metallization" forms the critical interconnections between different areas of the chip and the different transistors. Metallization is also used to form the bonding pads that connect the chip to something called the package and then to the circuit board of the system it supports.

At many places during the chip making process, the wafers are cleaned and polished, layer after layer in dry processes or wet processes until the wafer is finished. After all these automated production steps are complete, a final protective layer is put over the entire wafer.

Depending on the manufacturer, testing can occur at various stages of production. Some vendors use a probe test to provide an initial look at how many functional devices are on the wafer. At IBM's 300mm plant in East Fishkill, which we are about to verbally tour, Big Blue uses on-chip testing so the test circuits on the chip itself, talking to special testing computers determine one uncut chip at a time, whether a particular chip on the wafer has been built correctly.

Current technology allows chip vendors to pack more than 800 transistors onto an area as small as the diameter of a single human hair. At such tiny dimensions, as you would expect, even the smallest dust particle can ruin the functionality of an entire chip.

When the wafer is fully built, the wafers move to another section of the plant or to another plant where they are sliced and diced. A very precise saw cuts the individual chips from the wafer and the chips are then often referred to as a "die." They are further packaged, tested and tested again in a "final test" and they are then packaged for shipping and shipped to the ultimate customer.

Figure 3-5 Completed Wafer

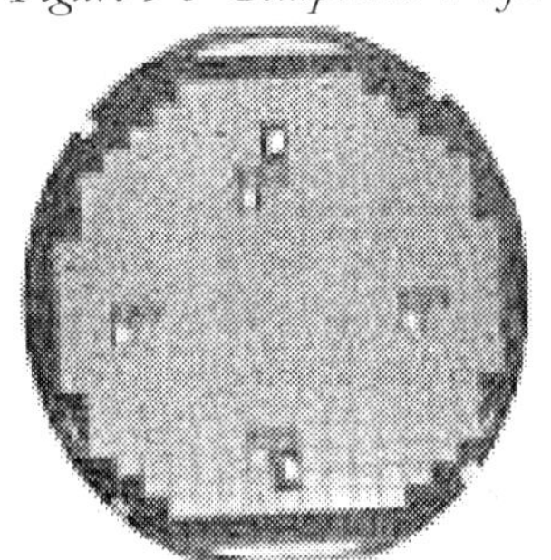

Although the above steps are in general use across the semiconductor industry, a company's proficiency in manufacturing its chip designs can be the difference between winning and losing in this highly competitive market.

These finished semiconductors eventually do all kinds of great and or mundane things, such as guiding spaceships and handling data by controlling the flow of electrical current to make words, numbers, sounds, images and colors. They are used worldwide by all major integrated circuit manufacturers and indirectly by nearly every human on the planet. Applications can be found in commonplace items like computers, telecommunications and televisions, as well as in more advanced applications like microwave transmissions, laser switching systems, medical diagnostic and treatment equipment, defense systems and of course, even the NASA space shuttle.

Precursor to the Plant Tour

Reading the technical papers, scouring the news services, and interviewing chip manufacturers are all good ways to get the information necessary to write a book about semiconductor chips. I've done all three and most of this book is based upon this research, but that's not all. There were some conferences, interviews and emails with some pretty knowledgeable people who helped me put all this chip material in focus.

In July 2005, I was invited to East Fishkill NY by Thomas Curran of the IBM's Analyst Relations Department. After a phenomenal plant tour, I got to meet Dr. Bernie Meyerson, IBM's Chief Semiconductor Technologist (CTO), an IBM Vice President and most significantly, an IBM Fellow-- IBM's top honor for a corporate scientist. Along the way, Dr. Meyerson has some significant inventions and patents to his credit including the use of a combination of silicon and germanium (SiGe) to produce very fast, low power chips that are used today in small devices like PDAs and cell phones.

In March 2005, at the COMMON Conference in Chicago, I got really interested in the ongoing "Chip Wars," after I had the pleasure of sitting in on a session taught by IBM's Dr. Frank Soltis. Dr. Soltis is IBM's eServer iSeries Chief Scientist and he is widely known in the industry as the father of the IBM AS/400. He also led the design of the 64-bit PowerPC processors used in the IBM iSeries and pSeries servers. He is quite a person and it was an honor to be in the same room.

When I came back from Chicago, I was so excited about what was happening in the chip industry, I began to research more and more about it. Dr. Soltis had introduced me to so many different notions about the current state of the semiconductor chip industry that I needed to learn more. As a writer, throughout this book as you have seen, I have been sharing my learning and my perspectives of this fascinating industry, and we are approaching the part of the book in which the topics are technical spicy, if not hot. After we take the plant verbal tour, the next three chapters are about Intel, AMD, and IBM as major players in the semiconductor industry whose mutual cooperation and competition is not always done in the spirit of coopetition. In fact, some might suggest that all three are engaged in a "Chip War."

In performing more and more research for this book, I had expected to visit Intel, AMD and IBM for the information that I needed. For the questions I posed, Intel and AMD referred me to the "Press" section of their Web sites and offered a phone and email dialog. That was about all the action I got from them but in all fairness I did get tons of information from their well crafted Web sites as well as all the stories that are now out about their very public war. IBM on the other hand, was so interested in this project that the company assigned one of its Analyst Relations people specifically to me.

By the time IBM had feed up enough time for me, I had already completed most of the book. However, after my plant tour, I did more detailed research, contacted Intel and AMD for more information, revisited the whole book, added a few chapters and corrected some incorrect assumptions on my part. Overall, the book is much better now and it is even more accurate.

The Verbal IBM Chip Plant Tour

Along the route to a book that was fit to print, my publisher, Joe McDonald and I took IBM up on its invitation its chip fab in East Fishkill NY. Before meeting Dr. Bernie Meyerson, IBM's CTO, we were first given a tour of its clean room 300mm, fully automated wafer manufacturing environment. We got our first sense of what clean was when we were asked by our guide prior to entering the "non clean" section of the plant if we would be kind enough to shine our shoes in this huge vacuum powered machine. The machine was so powerful that it was difficult just to get our feet out of the machine and back on the ground. On the tour we learned exactly what clean means to IBM. It is hard to believe, but IBM's manufacturing environment is ten times cleaner than that of a hospital operating room.

5X Is Not a Chip Model Number

I know I am not really that big but the coverall part of my clean room uniform was a 5X. We looked splendid in our body covering garb, from foot mittens to latex gloves to a hair net and a fancy head piece that came down just past the eyebrows. We both looked like nuns in a hospital. I looked especially interesting since I got to wear a special veil in front of my face, above my nose and below my eyes to protect all the IBM fab equipment from the potential of a flake from my mustache. The veil was required for anybody with facial hair. What a wonderful experience. It took about ten minutes to get ready.

Once inside the clean environment we quickly observed a cadre of upside down, independently operating R2D2 units running on train tracks. These units whisked plastic "boxes," called Front Open Uniform Packages (FOUPs, each containing as many as twenty-five 12" silicon wafers from automated workstation to automated workstation. I had to get closer for a better look at this process.

Figure 3-6 FOUP Being Snatched Back into R2D2

Because the plastic boxes were opaque, I could not actually see the 25 disks in the package as the FOUP (a little blurry in Figure 3-6 also) was being operated upon at a close-by workstation. Our tour guide, Raymond Knauss, a very knowledgeable project leader in one of the bays often began his introductory words at a new tour area with, "as you can see." In this particular case, I could not really see the 25 disks. So, to get a better look at the technology at one fully automated workstation, I moved in closer to the translucent window in front of the 25-chip container.

Bells went off like I had just won a jackpot at the casino and my hosts' eyes were popping from their heads. They had all they could do to not pull me back abruptly from destroying the day's production. Before I knew it, a long copper-like arm with big fingers came down from the R2D2 unit directly above the workstation and it pulled up the box and protected it from the human intruder, yours truly. When I retreated back to the safety area, the arm delivered the box back to the unit and production began again like clockwork. The picture in Figure 3-6 is not me but it is representative of the FOUP being snatched from the workstation and my looking on in amazement. Thankfully the R2D2 units were not armed with rubber bullets or mace and I emerged safely from the tour and I was not arrested. What a system! What an experience!

From a receiving area in which fully sealed cartons of 12" silicon substrates arrive from suppliers such as the Soitec Group (Euronext, Paris -- the world's leading supplier of silicon on insulator (SOI) and other engineered substrates) or Japan's Shin-Etsu Handotai Co. Ltd. (SEH, the world's largest wafer producer), or other wafer suppliers, the containers are last touched by human hands. They are started on their way on a journey that involves the many activities of chip production that we outlined above. In two months, they'll be off to Burlington, Vermont for slicing and dicing.

In about two months from start to finish, after all of their deposits, baths, and cleaning treatments, the 900 or so uncut chips on each wafer have grown to twenty or more layers thick and are ready for the trip to Burlington Vermont for final testing, cutting, and shipping to IBM's customers. The good news for IBM, the FOUPs, and the many R2D2 units who now have my picture in their non-destructible flash memory, as well as for me, is that I won't be getting shipped to Burlington along with them.

Chapter 4
Early Chip Pioneers

Shockley Semiconductor-- the Chip Pioneer's Pioneer

Out of a number of top level semiconductor chip manufacturing companies, Intel today is probably the most well known. As many youthful Intel chip users (PCs and servers) look at the names of the companies that Intel had to pass in order to become number one in its business, many of the names may be unfamiliar. Nonetheless, the companies were formidable in their heyday. Many of the names and faces of the pioneers are still with us, working at one major fab or another.

The first company on everybody's list of pioneers is Shockley Semiconductor. This chapter is both Shockley's story and the story of the beginning of the semiconductor (chip) industry.

When William Shockley was ready to leave Bell Labs to found his brainchild, Shockley Semiconductor in 1955, as one of the first companies in the newly forming semiconductor industry, he took with him a legacy of invention and eight of the most talented young men who would ever work anywhere and at any time in the computer field.

Shockley decided to form his company to build transistors. After all, eight years had passed since he, John Bardeen, and Walter H. Brattain had invented the transistor for Bell Labs. It was a time to celebrate at Bell Telephone Laboratories in December, 1947 with this historic introduction. But the celebration had long passed.

Figure 4-1 William Shockley

The Devil Made Me Do It!

Shockley was aware of all the industry buzz about computers and what they would mean in the future, and Bell Labs would become a pioneer in building computer components for their huge telephone switches. There were lots of highly technical people with whom to discuss ideas at Bell Labs since the company employed only the cream of the crop of graduating scientists and engineers. As the accomplishments increased and big time opportunities seemed to be available on the outside, despite what a great company Bell had, lots of Bell Labs employees were willing to jump ship for the chance to make it really big.

Before he made the jump Shockley knew that he needed some financial help to get through the startup phase of a new business. He received some funding from Beckman Industries and he formed his new company in Palo Alto California, which is near Stanford University, in the heart of what is now known as Silicon Valley. Shockley Semiconductor was thus the first semiconductor company to arrive in that area of California. In many ways William Shockley gave birth to Silicon Valley.

He pulled his group together, to get on with their work, not too long after leaving Bell Labs. The team with whom he worked is well known in the industry today, especially to computer scholars, and has often been referred to as the greatest collection of electronic geniuses ever assembled.

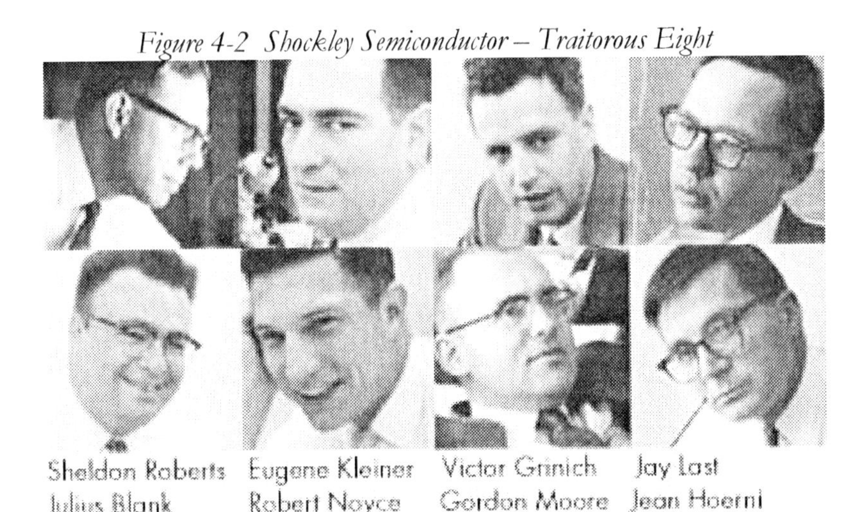

Figure 4-2 Shockley Semiconductor – Traitorous Eight

Good Things Do Not Always Last!

Unfortunately for Shockley Semiconductor, they would not be assembled for too long. His company did not last long. It never did make it big, and after just a few years, Shockley sold out to Clevite Transistor in April, 1960. While it may not have been a success financially, its industry impact was felt big time. Shockley himself was recognized as a genius as an engineer, but his management style was in conflict with the group. He drove eight of his key employees away within the first year.

The Founding Of Fairchild Semiconductor

This list of men who left Shockley sounds like who's who in the history of computers. These men, shown in Figure 4-2, who became known as "the traitorous eight," were: Robert Noyce, Gordon Moore, Sheldon Roberts, Eugene Kliner, Victor Grinich, Julius Blank, Jean Hoerni and Jay Last. They left Shockley semiconductor in 1957 and together formed Fairchild Semiconductor, which became an "independent" business unit of Fairchild Camera.

Fairchild Semiconductor did pretty well for itself after some major semiconductor innovations by Jean Hoerni were implemented in the 1959 time period. During Fairchild's heyday, the number of employees ballooned from a handful to about thirty thousand, and sales rocketed from a few thousand dollars to about 150 million dollars per year.

Fairchild Experts Go Elsewhere

As Fairchild and the semiconductor industry grew, many of Fairchild's original eight saw opportunities elsewhere. Two of the most notable, Moore and Noyce left Fairchild in 1968 to manage their small start-up company which they eventually named Intel (Integrated Electronics). Noyce was the general manager of the company and was responsible for Intel getting its chance to get inside.

Robert Noyce had left Philco Corporation (TVs et al.) where he had been able to perform research on leading edge electronics. Gordon Moore left his chemistry research position at Johns Hopkins University's Applied Physics Lab. They both set out to do something special with their lives. They had issues with Shockley, almost immediately, and chose to move out, along with the other members of the infamous "traitorous eight." Shockley was upset with the whole bunch of them leaving at the same time and the notion of a large scale mutiny is conjured up with his "traitorous eight" label.

Life Was Not That Bad

Moore and Noyce did very well for themselves at Fairchild Semiconductor. Noyce had become general manager, and Moore was head of Research and Development.

While at Fairchild, Noyce made semiconductor history by independently inventing the integrated chip. However, at the same time, Jack Kilby was involved in a similar project at TI, where in 1958 he also, independently, invented the integrated silicon chip and patented his submission. History credits both men with this invention.

Though Moore and Noyce were doing quite well at Fairchild, what they received was not exactly what they had bargained for. There were major changes in management at Fairchild Camera and strange things were happening in the business. The Fairchild Camera (the mother company) board of directors was diverting funds from the semiconductor group to areas outside the chip business. There was not enough money left in the chip company to continue innovation. Noyce and Moore were not pleased with the situation and they decided that they had enough. The rest of their story is called Intel and we have a whole chapter on that just ahead.

Other Pioneers Did Time at Fairchild

Besides the "traitorous eight" who left Shockley to form Fairchild Semiconductor, along with Sherman Fairchild himself, there are a number of other chip industry pioneers. They all toiled at Fairchild Semiconductor and got to know each other quite well. This list includes Robert Widlar, who was the first to engineer an analog IC, Jack Gifford, the CEO of Maxim Integrated Products, one of the silent cofounders of AMD, Andy Grove one of Intel's three co-founders, and Jerry Sanders who was the strength behind AMD for many years before his retirement in 2002.

Other Pioneering Chip Companies

Shockley Electronics and Fairchild Semiconductor were not the only ones doing work with semiconductors in the 1950's. Texas Instruments and Motorola were two big and better-known companies that got involved with integrated circuits early on and in a big way. Of course there was also Bell Labs. Among the lesser knowns of course were Shockley Electronic, Fairchild, MOS Technologies, Zilog, and then AMD. All of these companies combined, plus Intel and one of the least known major chip innovators of all time, IBM, are responsible for building a substantial portion of all the computer chips that have ever been made.

Some of these companies, such as TI and Motorola have some notoriety because every now and then they have been in the consumer marketplace with items such as radios, watches, TV's, stereos, and home computers. Others are well known because they are the companies who build the computer processors for Sun, IBM, Dell, HP, and Apple.

Wintel

Intel, of course has been in the public eye since the advent of the IBM PC in 1981 and of course because of the company's close relationship with Microsoft. This relationship has caused many in the industry to refer to this combination of computer chip vendor and OS vendor using the term "Wintel." When compared against other solutions, Wintel products appear homogenous and the duopoly appears to be a company by itself. In many ways it is. But with the advent of the public and silent chip wars, as you will see, the Wintel partnership may not be quite as solid as it once was. Time will tell.

IBM and Bell Labs

Some might suggest that IBM and the old Bell Labs (Lucent Technologies) do not even belong on the list of pioneering chip makers. Neither of these companies have had a major influence on the initial commercialization of semiconductors per se. Yet, they are both major technology pioneers and major producers. The reason they may not belong with the pioneering legends is that the fruits of their labor have typically been used internally and not offered for general sale. In the last ten years, however, IBM has changed from this background model and now is very much a major force in the semiconductor industry, though it still goes about its business "quietly."

When IBM and Bell Labs were doing their share of historical innovation – even before Intel, they used their discoveries to help gain the competitive edge in their respective markets. There was no semiconductor industry per se. Bell has a number of inventions to its credit including the transistor; and its signal and high speed phone switching technology always eclipsed all players in the switch market. Bell Labs was simply the best in its industry.

From the early 1960's leading up to the announcement of its famous System/360 family of computers, the IBM Company exploited silicon chip technology. Besides inventing the disk drive, IBM inventions contributed greatly to the expansion of the semiconductor industry. For example, IBM's Dr. Robert Dennard, in the early 1970's invented dynamic random access memory (DRAM). This is still the basic form of memory used in all modern computers.

IBM often outsourced its chip manufacturing. After seeing that many of its R&D secrets were frittered away as major subcontractors such as RCA and GE were emboldened to enter the mainframe business, the company chose to do its own chip manufacturing. So, for many years IBM has been a major but quiet developer and manufacturer of the powerful chips used in its proprietary systems, controllers, and devices. Until IBM broke from the mold in the early 1990's, neither IBM nor Bell Labs (Lucent Technologies) sold its chips commercially.

The microcomputer / semiconductor industry continues to be very exciting with many of the pioneers, including Intel's Gordon Moore, and AMD's Jerry Sanders still very much alive to share in their company's glory. In the next series of chapters, we take a longer look at a number of the companies that brought the semiconductor (chip) industry to where it is today. Let's start our adventure by first taking a look at the Texas Instruments Company.

Chapter 5
Micro Technology Pioneers: Texas Instruments

TI Has Affected Our Lives

Every now and then we hear about Texas Instruments doing something special which affects our lives. One of my first introductions to the company was when it tried its hand at digital watches in the 1970s. They were neat and very inexpensive. But they were not as neat as the TI/99 4A computer which could be bought for $50.00 at Boscov's Department Store in my home town along with many other retail outlets in 1982. The "99" package deal also provided a speech synthesizer that would work with many of the TI/99 game cartridges. There were tons of TI game cartridges back then or so it seemed, and they did not cost a "million dollars." The "99" was really quite a machine and it found its way into many living-rooms, long before Nintendo even began its play.

Figure 5-1 Ti99/4A Console with Speech Synthesizer

Courtesy of Jon's Ti99/4A Web Site www.guidry.org

Some of the game cartridges were not game cartridges at all. They were clever little programs that quickly made kids think that learning was fun. My son Brian thought his "99" and all of its learning tricks were lots of fun. Brian had a love affair with hand calculators from when he could barely speak. During the 1982 Christmas season, at 2 ½ he was speaking pretty well and he had become a whiz with his calculator. Before he had begun to speak well, his calculator was his "kuh-shah-shun." That's the phonetic spelling, of course. Eventually, he could say calculator, just like you and I, but mom and dad kept using "kuh-shah-shun." because we thought it was so cute.

Santa and the TI/99 4A

Santa gave the 2½ year-old a TI 99/4A for Christmas. It was complete with a special TI-invented speech synthesizer. Santa had set it up in the family room, connected it to the TV, turned it on, and had left a note for Brian by the tree so he knew where to find it. You know the look in his eyes when he saw it! Before we knew it, he was working one of the math cartridges for kids. He selected the hard problems. They started off with the machine speaking: "Two plus four plus one equals." When it finished it paused and in a deep, slow voice the synthesizer said, "Your turn!" Brian quickly "plugged-in" a "7" and the machine rewarded him with a series of happy sounding tunes and small fireworks images.

Then the machine said "twelve plus seventeen plus nine equals." Again, the machine said, in that same deep voice: "Your turn!" Brian thought for awhile and then when he was ready to "plug-in" the answer, he heard a buzzing and an unhappy sound and the machine read the problem again and gave him the answer... "38"... which, by then, he already knew! But he had not entered it in time. That would never happen again!

Brian paused the machine and he left the room. He came back in a minute or so with... you guessed it, the "kuh-shah-shun." The next time the machine said: "Thirteen plus fifteen plus seven," Brian was plugging the numbers into the "kuh-shah-shun" as it spoke. When the machine said: "Your turn" He plugged in the value "35" right from the display on the "kuh-shah-shun." He was never wrong again!

TI Products Are Likeable

I never met anybody who had a TI digital watch or one of their "99" computers who did not enjoy the experience. In the 1970s, TI also made complete minicomputer systems, but they chose to not stay in any of these three businesses for very long. The TI business computer users, who I met, were forced to move to another platform, but they liked TI and were moving only because TI was getting out of the business. TI chose not to be in the thick of the consumer business again. They knew how to compute but had a hard time with consumer and special marketing demands.

The Origin of Texas Instruments

TI has been around for a long time, going on eight decades, since May 16, 1930. They began as "**Geophysical Service**," an independent contractor specializing in the reflection seismograph method of oil exploration. The founders of TI include: J. Clarence "Doc" Karcher and Eugene McDermott. Cecil Green and Erik Jonsson were among their first employees. The company was incorporated in Delaware in December, 1938. In 1951 the company formally christened itself, "Texas Instruments."

TI And Technology

During this time TI has been a major innovator and it is well known to participants in the OEM digital and analog signal processor market. Many do not know however how much an impact TI has had upon the overall computing landscape. For example, it is not well known that TI designed the first commercial transistor radio in 1954, long before they were mass marketed in the US by Japanese companies. Moreover, it was TI precision switches, thermostats, transistors, and other semiconductor products which helped the Apollo mission land the first men on the moon. TI was on the moon even before Ralph Cramdon threatened to send Alice there!

Figure 5-2 Jack Kilby's First Integrated Circuit

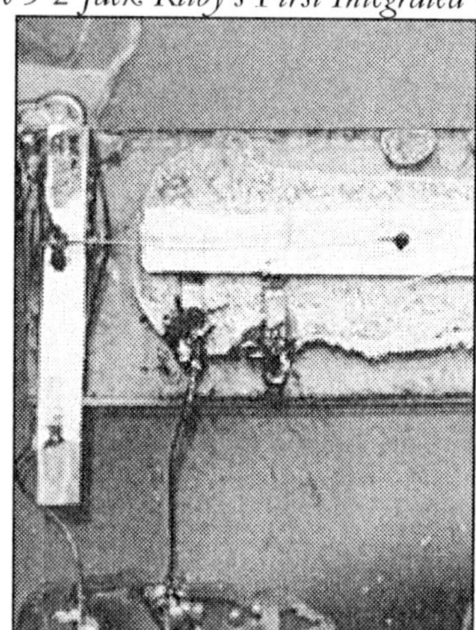

Courtesy of Texas Instruments, Inc.

TI has a number of other firsts to its credit, including the first commercial silicon transistor, the first integrated circuit as shown in Figure 5-2 (Jack Kilby of TI and Robert Noyce of Intel share the industry honor – Jack Kilby passed on in June 2005), the first microprocessor, the first single-chip microcomputer and the first electronic hand-held calculator. And for those who like to hear machines speak, they also invented the first single-chip speech synthesizer, the reason why everything from toys to greeting cards can talk to you and me. And, of course, the TI/99 voice was the motivation for my son to get out his calculator

Chapter 6 Micro Technology Pioneers: Motorola

Motorola the Chip Maker with Different Roots

Long before Gordon Moore, Robert Noyce, and Andy Grove began to build their fortunes at Intel, Motorola's founders were active building their company. At 33 and 29 respectively, Paul V. Galvin (1896-1959) and his brother Joseph E. Galvin (1899-1944) were a bit older than some of our modern day billionaires such as Jobs and Gates when they began their entrepreneurship by purchasing the battery eliminator business of the bankrupt Stewart Storage Battery Company in Chicago. On September 25, 1928, the Galvin brothers gave name to their enterprise as they incorporated the Galvin Manufacturing Corporation.

The First Product

The Galvin name was really no surprise as they quickly rid the company of all remnants of the Stewart Company they had purchased. But they did keep the five employees while renting new quarters on 847 West Harrison Street. Wages were not very steep back then, but even the $63.00 first week's payroll taxed the fledgling company. Their hard assets were meager, consisting of $565.00 in cash and $750.00 in tools so they had to work fast to create a successful product.

Fortunately, there was a prize hidden in this newly acquired company, the design for its first product, a battery eliminator. Today, we might call such a device an AC adapter. This clever device, when it was developed, enabled battery-operated home radios to operate on ordinary household electrical current. It wasn't too long before the company had to tackle its next big problem. As electronic tube radios took over for these battery units in households, battery eliminators appeared to be on their way to extinction.

Before having to tackle an obsolescence problem, however, the Galvin's were already on their way to their fortune. By 1930, after just two years, their net annual sales rocketed to over a quarter million dollars. At the same time the company diversified and built the first practical and affordable automobile radio. Since radios were not an option provided with an automobile at the time, the Galvin's had an instant success on their hands. Way back then, Paul Galvin recognized some value in a future company name change and he coined the term Motorola for the company's new product line. Motorola brings with it the connotation of both motion (motor car) and music (Victrola). As you might have expected, it was not long before Galvin's share of the auto radio business grew so rapidly that it established the company as a leader in the U.S. market.

Figure 6-1 Motorola in the Car Radio Business

From Computer Desktop Encyclopedia
© 1996 Motorola, Inc.

(Image courtesy of Motorola Museum of Electronics.)

New Products, Nice Growth

Product line innovations came fast as the company invented the "Police Cruiser", an AM auto radio pre-set to receive police broadcasts. This was Galvin's first entry into the new field of radio communication products. Of course today the company is a leader in pager products and many wireless computer devices.

By 1940, net sales were almost $10 million and the employee population had grown to almost a thousand, as Galvin Manufacturing became a pioneer in FM radio communications. During the same period, the company developed the first hand-held two-way radio for the U.S. Army Signal Corps. The portable "Handie-Talkie" two-way AM radio quickly became a World War II symbol.

The War Effort

Just like all of the other electronics companies and business machine companies of the day, Galvin committed its full resources to the war effort. Once the war broke out, the company did sell 125,000 already built radio units for home use. But, they were the last shipped until the company went back to peace time production in 1945. It certainly must have been hard to get a radio back then.

New Name, New Directions for Galvin Manufacturing

By 1947 when the company offered its first public stock offering at $8.50 per share, the Motorola trademark was so widely known that Paul Galvin changed the company's name to Motorola, Inc. By 1948, as IBM and Univac were duking it out for computer champion of the world, Motorola made its initial entry into the TV business with its $179.00 "Golden View" offering. Considering that black and white TVs currently sell for less than $100.00, we can get a sense of how privileged one had to be in 1948 to be able to buy one of these. On top of that, Motorola's seven inch "Golden View" was the least expensive TV of its time.

Figure 6-2 Motorola Golden View 7" TV

While computers were still being built with tubes, in 1952, Motorola began a semiconductor development group. Their first efforts were in the manufacture of a 3-amp power transistor. Being a technology developer and a product manufacturer, the company eventually used its technology to build a commercial product. The auto radio which incorporated transistors for the first time was introduced in 1956. By 1959, Motorola technology had gotten so good that their new, smaller and more durable, all-transistor auto radio quickly became considered as the most reliable in the industry.

After a while, Motorola got good at making components as well as consumer products. Though initially, the company manufactured transistors and other semiconductors for exclusive use in its own radios, televisions, and communications products, in 1959 Robert Galvin, son of founder Paul Galvin decided to reduce its costs of component production by becoming a commercial producer and a supplier of semiconductors for sale to other manufacturers. Motorola was officially in the semiconductor business.

By 1960, net sales had climbed to just about $300,000,000 and the employee count was approaching 15,000. The company became a leader in all facets of the electronics world. One of Motorola's chief customers during this period was the US Government as the company built a transponder for the Mariner II for its flight to Venus. This unit provided a radio link spanning 54 million miles. When Mariner IV hit the sky in 1964, Motorola was there again with a transponder used to send images of Mars all the way back to earth. Motorola remains a major supplier in the space program and has been active from the Gemini Program through the Hubble Telescope to many top secret efforts underway today.

Many Products

Like Texas Instruments, Motorola has been in all aspects of the electronics industry from semiconductors to home products to space equipment. The company also likes to highlight its consumer product accomplishments on its Web site. In the late 1990's if you took a quick look, you would have found the Motorola name associated with many diverse products such as the following:

- ✓ Rectangular picture tubes
- ✓ 8-track tape players
- ✓ "Pageboy" radio pager
- ✓ Low cost plastic encapsulated transistors
- ✓ "Quasar" line of color receivers,
- ✓ (America's first all transistor color television sets)
- ✓ Portable FM two-way radio "Handie Talkie"
- ✓ Components for battery powered quartz watches
- ✓ Integrated circuits, quartz crystals, and miniature motors
- ✓ Short-range radio frequency radio telephone system
- ✓ Electronic engine control modules
- ✓ Pocket Bell pagers
- ✓ Instrumentation for cars and trucks
- ✓ Control modules for engine transmissions.
- ✓ Secure telephone terminal
- ✓ "Micro-TAC" personal cellular phone
- ✓ Lightweight Satellite Terminal (LST) radio
- ✓ Wireless In-Building Network (WIN)
- ✓ etc.

Today, Motorola's product range is even more diverse and the company still is in both the consumer marketplace and the component marketplace. You'll find a vast array of hardware products for example listed on Motorola's Web site under the following headings:

- ✓ Broadband Peripherals & Accessories
- ✓ Cable Modems & Gateways
- ✓ Cordless Phones
- ✓ Home Entertainment
- ✓ Home Monitoring & Control
- ✓ Home Networking
- ✓ MP3 Players

High Technology

Though the company originally became well known for its consumer products, over the years, Motorola has moved out of a number of consumer areas to focus on their main interests - technology. For example, in 1974, it pulled out of the TV business, selling the Quasar brand to Matsushita Electric. In 1987 Motorola produced its last auto radio. At the same time, the company also divested itself of its display systems business as well as its automotive alternator and electro-mechanical meter product lines.

Motorola has also acquired a few companies along the way to strengthen its presence in technology areas in which it wanted to better participate. These include Codex Corporation, a leading manufacturer of products and systems for data communications networks, and Universal Data Systems, a maker of moderately priced data communications equipment.

The reason Motorola is in this book is not because of most of the items on its product list. It is because they were and continue to be a major player in the semiconductor arena. Hundreds of companies including Apple have used Motorola processors and semiconductor components in thousands of products from coffee makers to PCs to controllers for mainframe computers.

As Motorola became interested in selling its components, the company chose to launch a new subsidiary. Founded in 1953 as the semiconductor products sector of Motorola, Inc, Freescale quickly became a leading global semiconductor company focused on providing embedded processing and connectivity products to large, high-growth markets. Freescale currently focuses on providing products to the automotive, networking and wireless communications industries.

Motorola's Microprocessors

Motorola / Freescale have been in the microprocessor business for a long time, having introduced its first microprocessor, the 6800 in 1974. This unit contained approximately 4,000 transistors. The specs at the time were unbelievable as the chip needed only a single 6-volt power supply and was supported by a range of RAMs and ROMs, and other devices, thus making it an ideal chip to select for mid 1970 era computers or controllers.

The early 6800 customers came from various heavy industries, such as automotive, communications, industrial, and business-machines sectors. It is also a fact that Steve Wozniak, one of Apple's founders, loved the 6800 so much that he wanted to make it the engine for the Apple I. However, as the story goes, the "Woz" was a great shopper and he found the big blue special light on at MOS technologies, a Motorola break-away company. MOS was charging just $25.00 per 6502 copycat chip vs. Motorola's $125.00 for the original 6800. Though one might say that the 6800 was an inspiration for the Apple, none of these chips found their way into the Apple PCs of these early days.

Figure 6-3 Motorola MC68000 Module

Moving from technology generation to generation, in 1979, two years before IBM's PC was introduced, Motorola introduced its first 16-bit microprocessor, the 68000. This was a hummer of a machine and its power and facility was great enough that Apple would begin a relationship with Motorola that continues to this day with the 68000 becoming the base microprocessor for the Macintosh. The 68000 was capable of completing two million calculations per second, and Motorola touted its architecture as enabling the unit to be used both to run and to write programs for scientific, data processing, and business applications. The processor became the mainstay of the Macintosh. In 1984, Motorola upped the ante again with the first true 32-bit microprocessor dubbed the MC68020. This unit had over 200,000 transistors with access up to four billion bytes of memory. By 1986, more than 125 companies were producing systems that used this processor.

For its own reasons, even though the Motorola 8-bit 6800 chip was available in 1981 when IBM launched its PC, Big Blue chose not to use the Motorola unit. About ten years later, having become a major chip foundry and looking for some chip business of its own, IBM representatives made a sales call on Apple. IBM offered the PowerPC RISC (reduced instruction set computing) processor as an option for Apple's next generation of computing. Apple informed IBM that it did its chip business with Motorola. Undaunted by the setback, IBM packed up and went over and made a sales call on Motorola. The result of course was that IBM, Apple and Motorola formed a partnership on the PowerPC chip line.

For years Motorola, through its Freescale Division and IBM made PowerPCs and Apple and IBM continue to use them in their computers and controllers. Though Motorola partnered with IBM in its PowerPC chips, the company still designed and produced its own proprietary 88000+ RISC base and 68000+ base follow-on chips for a brief time. Motorola's Freescale Division got out of the PowerPC chip business in 2004.

Displaying its penchant for high quality internal processes, in 1988 Motorola was declared the winner of the first of the very prestigious Malcolm Baldridge National Quality Awards, given by Congress to recognize and inspire the pursuit of quality in American business.

Meteoric Business Growth

During the eighties and nineties, business was booming and, despite no longer making radio noises, the company continued to reap in the cash. For example, in 1980 net sales zoomed past the $3 billion mark for the first time, more than tripling to just under $11 billion in 1990. With another ten years under its belt, Motorola took jackpot again with reported earnings of over $30 billion in 1999. During this expansive period, the number of employees also grew by more than 50,000.

In 1996, Christopher B. Galvin took the reins of the company as its Chairman and Chief Executive Officer. Galvin has that familiar name from the 1928 founding of the company. He is the grandson of Paul V. Galvin, and the son of Robert W. Galvin who had succeeded Paul Galvin in 1959. Though Motorola is a publicly traded company, it still is very much "galvanized" behind its founders' family

When you consider just how successful Motorola under the Galvins has been, and how long they have been reaping the rewards of their success, don't you wonder why the name Galvin is not better known. As a quiet company, Motorola lets its products do the talking. However, if we peeked into the assets of the Galvins, we would have to find net worth values in the neighborhood of the other personal computer billionaires.

Chapter 7
Micro Technology Pioneers: Zilog

Zilog: The Original Clone Company

W.C. Fields, when asked if he croqueted in one of his famous movies, once scoffed that he had been the champion of the Tri-State League and the Lesser Antilles. Then as he walked forward and fell, he questioned: "what lazy lout left all these wires on the lawn." Unlike Fields, Zilog was once the chip champion, but it was more than little wires that got in their way. When all the other chipmakers were forming their businesses, Zilog was claiming victory after victory. Many of us who went through both the minicomputer revolution and the microcomputer revolution can recall the Z80 as a formidable competitor for the pleasures of the microcomputer world. As a champion, the Z80 microprocessor is the undisputed largest-selling 8-bit chip of all time.

Z80 Standard

In many ways, the Z80 chip and an operating system from a company formed by Dr. Gary Kildall, known as Digital Research, along with a bus known as the S100 were the standard in "personal" or "hobby" computing in the mid to late 1970's. Most of the companies who engaged in the home computer business were startups, but one company took to the Z80 chip big-time. The Z80 was chosen to be the main engine for the Radio Shack TRS-80 line of computers. Using the Z80 as its main processor, the TRS-80 became very successful in the late 1970's and was Apple's major competitor. Radio Shack was so close to the notion of the Z80, that they named their prime machine after it - the Tandy Radio Shack -80, a.k.a., the TRS-80.

When Apple came out with its successful Apple II computer, Steven Jobs was seen scrounging around for some software to help buoy the unit sales. To this end, Bill Gates' fledgling Microsoft organization led the efforts by building a "softcard' containing a Z80 processor and memory. When an Apple II was equipped with one of these Microsoft softcards, all of the software written for the Z80 microprocessor was immediately usable on an Apple II machine. This was a technological coup for Microsoft, a marketing coup for Apple, and a big winner for Zilog as it strengthened its place in the early chip race. Despite these successes, IBM's decision to go to Intel in 1981 would change the direction of personal computing for all time.

Where Did Zilog Come From?

Zilog was founded in 1974 by Federico Faggin, one of the inventors of the microchip at Intel. He brought Ralph Ungermann, one of the managers directly reporting to him at Intel along with him as he started Zilog.

Faggin is definitely a member of the "whose who in chip pioneering." In fact, he is also a Fairchild Semiconductor alumnus, starting his career there in 1968 after having done leading edge work with Olivetti in Italy at age 19 and after having studied at the University of Padua. Following a short stint at Fairchild, Faggin was lured to Intel to work on a contract in an effort to produce the first programmable digital computer chip.

Though Ted Hoff of Intel is credited with the creation of the first microcomputer at Intel, many on the inside know that it was Faggin's work along with the main engineer Masatoshi Shima from Busicom, Intel's customer, who did the bull-work to innovate the chip. Because Faggin left Intel, there is major speculation that Intel chose one of their own to pass the full credit to. Yet, if you look at the first computer chip, in two different forms, clearly on the chip itself are the initials "FF" and those initials are lots different from "TH." The two chips are shown in Figures 7-2 and 7-3 respectively.

Figure 7-1 Federico Faggin's Initials Inside Intel 4004 Chip

Figure 7-2 Federico Faggin's Initials Outside Intel 4004 Chip

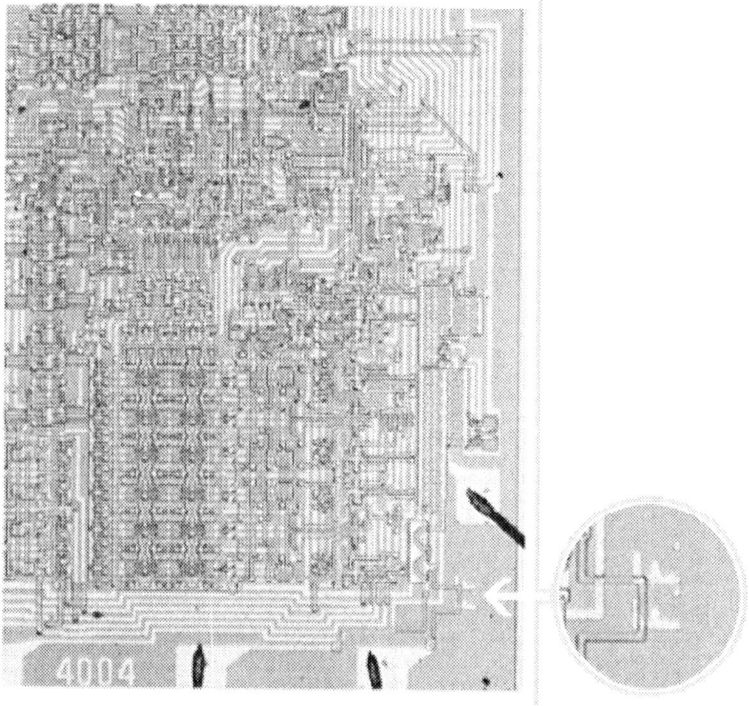

Faggin left Intel and used his resources along with Ralph Ungermann and others to launch the company with the Z80 chip as its main product. Faggin conceived the Z80 microprocessor family in December 1974. It was a Saturday. He tells a story of how he had been struggling for some time to figure out the company's first product. Then it came to him in a flash. Fagin liked to tell the story that when the idea came to him he said aloud, "Super-80!" With that exclamation, the basic architecture of the Z80 family came vividly into focus. The Z80 is still in high-volume production today, 30 years after its market debut.

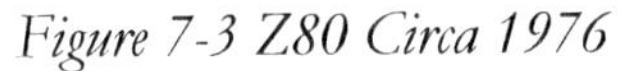

Figure 7-3 Z80 Circa 1976

Zilog today concentrates on the 8-bit simple microcontroller marketplace and has matured its family of processors with a number of products that are improved take-offs of Faggin's famous Z80. You don't find Z80's in personal computers or hobby machines anymore. In fact, for most of the company's history, its devices have been used in the back room, out of site from the logo-seeking public. The type of chips in which Zilog actually excels are humbly found on microcontrollers, well out of sight and under the covers of the actual piece of equipment which is being controlled. These are chips that go into things such as television sets, remote controls and similar electronic products.

If it were not for IBM's selection of Intel's 8088, the popular 8-bit Z80 chip could have helped spur spectacular growth for Zilog. It would have been Federico Faggin instead of Gordon Moore, Robert Noyce, and Andy Grove who would be walking the Brinks trucks to the bank. However, this did not happen. Though growth was steady, it was not raucous and there was always a problem area with which the company had to deal. Major competitors such as Intel, Motorola, Texas Instruments and Mitsubishi were too powerful for Zilog chips to make it in the consumer area but the company has always been successful doing its thing under the radar.

Selling the Company

Faggin himself did not keep the company very long, selling out to Exxon in 1981. He then co-founded Cygnet Technologies, Inc. in 1982 and served as President and CEO. Faggin also co-founded Synaptics, Inc. in 1986, where he is currently chairman of the board. He served as a director, president and CEO from 1987 to 1998. Since 2003, Federico Faggin is president and CEO of Fovoen, Inc., a start-up company dedicated to developing and marketing advanced CMOS image sensors.

Faggin is the recipient of many honors and awards including the 1988 International Marconi Fellowship Award, the 1994 IEEE W. Wallace McDowell Award, and the 1997 Kyoto Prize. In addition, in 1996, Faggin was inducted in the National Inventor's Hall of Fame for the co-invention of the microprocessor.

Zilog Post Faggin

Exxon later LBO'd the company to Warburg Pincus, which brought the company public again in 1991. For a while in the 1990's Zilog began to do well again as revenues grew at a nice 18% per year. During this period, the company diversified into modems for satellites and they formulated a marketing strategy of going after relatively small semiconductor custom orders. Never on top of the pile for too long, the Zilog Company quickly found more bad luck. By 1997, they were barely avoiding red ink on $261 million of sales. By this time, the stock had slumped to less than $20. The death spiral had to end or it would take them under.

After these years of unexceptional performance, the company still had survived and was in tact well enough to be acquired in 1997 for $527 million by the Texas Pacific Group, who after a brief period, took the company private again. This acquisition breathed enough life into Zilog for it to make a full recovery and to be able to grow again. In 1998, the newly energized Zilog Company hired Curtis Crawford, who had been Lucent Technologies Inc.'s microelectronics group president. As its CEO, Crawford was brought on to move the ship towards successful waters. Today, the company's efforts are focused on embedded systems, communications and home electronics devices. In 1999, Zilog began expanding again with the acquisition of another semiconductor company Seattle Silicon Corp.

Chapter 8 Micro Technology Pioneers: MOS Technologies

Chuck Peddle and Steve Sherman

From Bell Labs to Shockley to Fairchild to Intel, during the chip pioneering years, there were always great people who helped companies prosper and then would leave, start their own firms, and seek to prosper themselves. Motorola had its own defections over time and two of the more famous were Chuck Peddle and Steve Sherman. These two and a number of other Motorola Engineers jumped ship and formed MOS Technologies.

Chuck Peddle was an engineer who ultimately became Jack Tramiel's right hand man at Commodore. Steve Sherman eventually became Senior Vice President of Fairchild Semiconductor. The work which these folks inspired continues to this day on a myriad of 6502 microprocessor enthusiasts' web sites. 6502 units are still being made by various companies and the cult of individuals sponsoring this love affair has not been dissuaded to jump to the Intel platform.

As the story goes, in 1974 MOS Technology got its start and shortly afterwards it released a brand new microprocessor called the 6501. The microprocessor was so similar to the Motorola 6800 that Motorola cried foul. The chip was fully pin-compatible with the 6800, and though all of the user community was thrilled with its price, the price did not include the R&D, which Motorola had expended to design and develop the chip. Soon, MOS Technology was in court facing a lawsuit from Motorola. To have their cake and give Motorola's theirs too, MOS changed a few pin functions and by the spring 1975, they were free and clear to market their new 6502 chip which was just about all the 6800 ever hoped to be.

The unit quickly attracted the masses, not as much for its facilities, but for its price. MOS had developed a process to fix the photolithography mask to remove design flaws post facto. This made the company's yields about 70% compared to Motorola and Intel's 30%. The 6502 chips were not only loved by the masses, but the corporations quickly got into the love affair. Soon the 6502 began to be found in video game cartridges and video players as well as some top flight home computers from Apple, Atari, Commodore, and Nintendo. As clones go, they were as good as Motorola chips, at only 20% of the cost.

Figure 8-1 Apple I Circa 1976

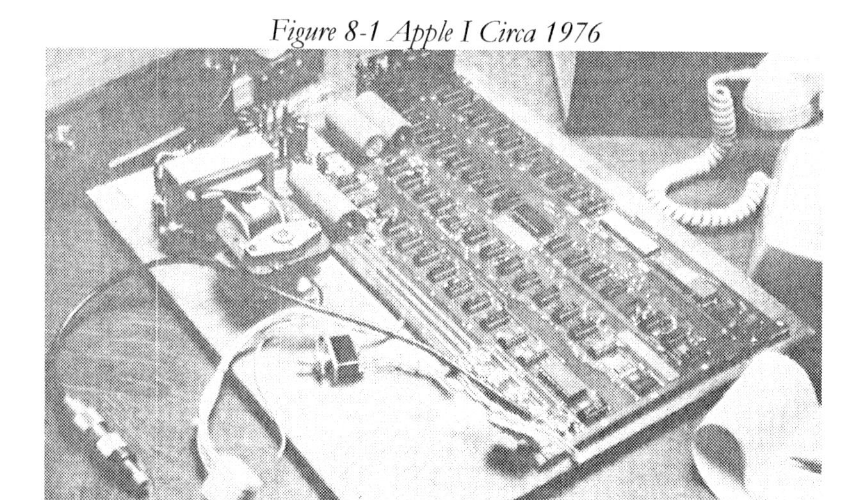

Just In Time for an Apple

In 1976, Apple (through the design efforts of Steve Wozniak, Apple co-founder) put together their first computer. Approximately 200 were manufactured but, as you can see in Figure 8-1, they were just boards with IC's attached. Each owner had to buy a keyboard and power supply, then build a case to house the "complete" computer. They called the unit the Apple I, and, with Wozniak's shopping talents, they chose the MOS Technologies' 6502 Processor.

As you can see in the picture, their Apple I was not much more than a single-board computer for hobbyists. However, along with the 6502, the unit also had a built-in video interface and in-board ROM that was used to get the machine fired up so that it could run other programs from external sources. In 1977 Apple put a nice case on the Apple I, fixed some problems, added some more goodies like a color adapter, and introduced their very successful Apple II line of computers – shown in Figure 8-2. As you would probably bet and be correct, the Apple II came equipped with the same 6502 processor as the Apple I. Both Apple and MOS technologies were on their way to success.

Figure 8-2 Apple II Plus Unit Circa 1980

Unlike the 8080 and Intel's bulkier offerings, the 6502 (and 6800) had very few registers. It was not much more than an 8 bit processor, with a 16-bit address bus. Inside were one 8 bit data register, two 8 bit index registers, and an 8 bit stack pointer. The 6502 was a technician's dream. One of its salient points was that there were actual secrets built into the machine. In other words, there were some undocumented instructions which the wise could discover and have lots of fun with. This made the 6502 a techie's dream processor.

While Apple started its love affair with Motorola ala the MOS 6502, there was another home consumer oriented computer company stirring in the 70's with products such as the PET, the Vic-20 and the '64. Commodore would become a big company before it ultimately fell by the wayside in the mid 1990's. Recognizing the potential of the

6502 chip, Commodore bought MOS Technologies in the mid to late 1970's. Though the company really was not in the same market as Apple, it was interesting for Apple to have chosen a chip that would become controlled by a potential competitor.

Eventually Commodore ran into a bunch of issues, mostly of its own design, and the company became convinced that it was a home computer company and not a chip manufacturer. So, Commodore abandoned its MOS Technologies business as it focused primarily on its prime market. Apple, for its part began to purchase its next generation more powerful and more expensive CPUs from Motorola.

Bust Time in the Bahamas!

To finish the MOS / Commodore story, it is interesting to note that eventually Commodore, after having reached a billion dollars in sales, could not sustain its product line or its corporate life. After changing the corporate headquarters to the Bahamas, for tax purposes, the big Bahamas Mama went bust in 1995, along with it Commodore's control of the 6502 chip and its follow-on devices. Though there were attempts to resuscitate the company and its Amiga PC line, these mostly failed. Commodore et al is now well decomposed in the bone-yard of computers and the bone-yard of business.

The Amiga Survives

Though Commodore is long in the rotting, the Amiga has proven to have a life of its own. After being tossed around from company to company for five years or more, in and out of the hands of such stalwarts as Gateway Computer and others, Amiga again resurfaced in the late 1990's in two forms - hardware and software. The hardware version, called Amiga One is available through various hardware manufacturers. The major source is a UK company called Eyetech who offer Amiga compatibility using the PowerPC chip. The Amiga software version is designed to run natively on machines or even under control of other OS platforms such as Linux. This is being championed by a reasonably new organization known as Amiga Inc. It almost makes me want to cheer!

Chapter 9 The Indisputable Industry Chip Leader: Intel

The Founding of Intel

When Gordon Moore and Robert Noyce could not control the company they started at Fairchild Semiconductor, they decided to do it all again. They put together a business plan for Intel (rumor has it Noyce typed it on a single sheet of paper) and they went off to conquer the world. Their objective was to make complex integrated circuits. That was what they did best. They wanted to make something complex, which could have universal applicability. They each put in $250,000, and they received $2,500,000 of investment money to get the company off the ground.

In addition to creating Intel, Noyce, Moore, and And Grove who joined the company shortly after its founding, were accomplished scientists with many achievements under their belts. They made their mark in history in other ways than just being superior businessmen. Robert Noyce, of course, is credited with the invention of the integrated silicon chip. Noyce's was nicknamed the "Mayor of Silicon Valley." He was one of the very first scientists to work in the area -- long before the stretch of California had earned the Silicon name. Among other things, Gordon Moore is well known for having made an often quoted prediction regarding semiconductor evolution known as "Moore's Law." Andy Grove is no slouch either. He holds numerous semiconductor patents, has written over 40 technical white papers and he often shares his knowledge with students in a University setting.

Moore's Law

It isn't everybody who gets a "law" named after them. Having worked in the semiconductor business for years, while at Fairchild, Gordon Moore took notice to the fact that the density of computer chips was increasing so regularly that you can set your watch by it. You can predict it accurately. He observed that microchips were doubling in circuit density every year or so (twelve to eighteen months).

This recognized geometric growth in circuit density achievements over time has been "canonized" as a law known as "Moore's Law." Though many have associated a corollary to Moore's Law that computational power also increases at the same rate, this is not what Moore permitted to be canonized as his "Law." Moore's Law actually reflects the engineer's observed continual miniaturization of circuits. The performance corollaries were not Moore's doing. However, because computer power and the reductions in space required for transistors were proportional for forty or so years, future predictions of density computer power were simply extracted from past accomplishments in miniaturization techniques.

Moore's Law and its implied corollaries were almost always right on the money.... a smooth curve on a graph. Though Moore expected his law to be in force for just ten years, it seems that human accomplishment in the field of computing is still buoyed by it, or constrained by it, whichever. Until very recently both curves have not strayed far from the law.

Moore recognized that physical space will ultimately determine how long increases in densities can be achieved. After all, from his point of view, the law expired in 1975. But, as the separation between circuits gets to within atoms, without a technology that goes even deeper than atoms, even Moore's law is expected to expire some day. Right now, from a circuit density perspective, nobody knows when that will be.

In this 21st century, however, even while Moore's Law continues to bring forth density breakthroughs, with the smaller circuits and the increased power driving the higher performance, a new limiting factor has emerged. In a word, it is *heat*. As you will see as we recall the last few years, heat, or more accurately, heat dissipation has become a major limiting factor in bringing the continued fruits of the corollaries of Moore's Law to market.

Intel Management History

Intel was officially founded by Gordon Moore and Robert Noyce on July 18, 1968. At the time, its objective was not to become the leader in microprocessors. In fact, microprocessors were not even on the back burner. Instead, Intel's objective was to make semiconductor memory practical. At the time, computer memory used something called magnetic cores. In essence core memory was implemented using a matrix of small wires with ferric oxide coated magnetic "donuts." Depending on which direction current was flowing in the donut, that particular memory "bit" was either on or off and thus rendered a binary value of one or zero. When Intel chose to make memory practical in silicon, it was a far stretch since such memory implementation had already proven to be at least 100 times more expensive than magnetic core memory, the leading technology at the time.

Intel's founders felt that semiconductor memory's obvious advantages -- smaller size, greater performance, reduced energy consumption -- would win the day. From their perspective they needed to convince manufacturers to try the new technology. They not only did that, they invented the single chip microprocessor along the way.

When Moore and Noyce left Fairchild Semiconductor and put a wad of their own money into Intel, they had some substantial financial help from a venture capitalist named Arthur Rock. Rock put in $10,000 of his own money and raised another $2.5 million for the fledgling company. For his efforts, Rock became Intel's first official Chairman. Noyce became the CEO and Moore became executive vice president.

To be exact, Moore and Noyce actually did not call their company Intel, instead they gave it the name NM Electronics (Noyce / Moore). Later in 1968, however, they purchased the rights to the name Intel for $15,000 from a company that had been doing business as Intelco. It was not until 1969, however that the company launched its first product, the 3101 Schottky bipolar random access memory (RAM). Later that year the company also introduced the world's first metal oxide semiconductor (MOS) static RAM, which they named the 1101. With Rock at the chair and Noyce as the CEO, the company was off to a fine start.

In 1971 Intel went public with an offering price of $23.50 per share. The offering was successful and the company raised $6.5 million. In 1975, Robert Noyce was named Chairman; Gordon Moore, President; and Andy Grove was made executive vice president. In 1979 Intel was playing musical chairs again with its infamous troika. Gordon Moore took on the role of Chairman and CEO, Robert Noyce became Vice Chairman and Andy Grove was named president and chief operating officer.

IBM Helps Intel

In 1985 a few years after IBM bought 12% of Intel to help the company remain independent after it hit some hard financial times, Intel exited its original business of memory chips called DRAM at the time to concentrate its efforts on microprocessor technology. IBM sold its Intel shares and as promised closed its stake in the company as Intel returned to profitability in 1987. To celebrate its new profitability, Intel held worldwide "back in the black parties." The company closed 1987 with record revenue and net income.

1987 brought more changes to the executive troika. Andy Grove was named Intel President and CEO while Gordon Moore remained as chairman and Robert Noyce kept his vice chair spot. In recognition for his role in semiconductor and microprocessor history, President Reagan awarded Robert Noyce with the National Medal of technology. In 1988 Noyce became president and CEO of Sematech, a consortium of Intel and other high tech companies aimed at keeping the U.S. at the forefront of semiconductor manufacturing research

In 1990, Intel was shocked along with the rest of the semiconductor industry to find that Robert Noyce had died suddenly from a heart attack. Noyce held 16 patents for semiconductor devices, methods, and structures at the time of his death.

When Robert Noyce passed on, Craig Barrett became executive vice president of Intel joining Chairman Gordon Moore and President and CEO Andy Grove in the executive office. Recognizing his pioneering work in the semiconductor field, in 1990, President Bush awarded Gordon Moore the National Medal of Technology.

Intel Inside

In 1991, Intel Management launched one of the most successful marketing notions of all time, "Intel Inside." Considering that Intel products were not something that would be marketed directly to the public, some experts might consider this campaign superfluous since it would theoretically not affect direct sales. However, it was ingenious and highly effective as consumer after consumer wanted their new PC to have "Intel Inside."

Intel believed in the notion that the more brand recognition the company could get, the more likely the Intel brand would get preference when Intel based PCs and other products were in head to head battles with the competition. As it turns out, the pro-football couch potato on Sunday again becomes the CEO on Monday and again influences computer decisions. The "Intel Inside" campaign pays continual dividends for Intel to this day.

In 1993, the new troika played more musical chairs. Craig Barrett was named Intel's executive vice president and chief operating officer. Gordon Moore remained chairman and Andy Grove remained as President and Chief Executive Officer. Just two years after the kickoff of the "Intel Inside" campaign, besides financial rewards, Intel began receiving accolades noting the major impact of the campaign. Financial World ranked the Intel brand as the third most valuable in the world.

Andy Grove was named "Man of the Year" by Time Magazine in 1997. Intel became recognized as one of the most innovative and successful corporations of all time. Besides Andy Grove's 1997 award, the leadership positions at Intel changed again. Craig Barrett became president and CEO, Andy Grove became chairman, and Gordon Moore became chairman emeritus.

In 2001, Gordon Moore formally retired from Intel's Board of Directors. In 2002, Paul Otellini joined the executive suite as president and chief operating officer. Barrett remained the CEO and Andy Grove stayed as chairman.

Otellini Inside

On May 16, 2005, Intel again made major changes to its executive suite. The company anointed Paul Otellini, then 54, who had been Intel's president and chief operating officer, as Intel's chief executive officer at the annual shareholder meeting. He succeeded Craig Barrett, 65, who was named chairman after Intel co-founder Andy Grove, 68, stepped down from the post. Unlike IBM's recommended exodus age of 60 for chief executives, Intel CEOs have routinely stepped down when they turned 65.

So, now, as the biggest battle that Intel has faced since its early 1980's survival fight is heating up, the company is prepared to fight both AMD and perhaps even IBM head on with its new executive lineup. One thing is for sure, Mr. Otellini and company have a major battle on their hands.

In the next section of this chapter we cover the products that originally made Intel successful and those that the company hopes will carry it off to a successful future.

Memory First at Intel

When Robert Noyce, Gordon Moore, and Andy Grove brought Intel on-line, they were building memory semiconductors for the industry. In fact, Intel today continues as a leader in producing flash memory for the smallest of devices. When Intel took its first steps, however, the company wanted to do more than memory, but they needed customers for these yet-to-be wares.

At the time, electronic calculators were coming into being and replacing the old electro-mechanical models. Hoping to leverage its semiconductor position in the calculator industry, Intel tried to find an established calculator company with which to partner. However, it seemed like all calculator companies had already teamed up with an established semiconductor partner for their future lines. Eventually, Intel found a company called Busicom, a Japanese firm, just starting in the calculator business.

Busicom wanted to build both business and scientific calculators. The company had already designed the 13 chips which they wanted fabricated by Intel. These chips would form the basis of 13 different calculator models. The Intel engineers at the time were very busy making memory circuits to keep the company's cash flow moving and the company really did not have the engineering resources to take the Busicom design and build all 13 unique chips. But Intel really wanted and needed this business.

Why Not Build A Computer?

As they examined the issue, one of the Intel engineers, Ted Hoff, burrowed through the design and looked at what they were really trying to accomplish. After he understood the objectives, Hoff suggested to Moore, Noyce, Grove, and others at Intel, that all of Busicom's objectives for the 13 different calculators could be achieved with one general-purpose computer architecture, rather than thirteen unique solutions.

Hoff knew he was on to something good. When he was discussing the potential of this undertaking, he told Gordon Moore that he could see this type of chip being able to be a major component in such things as elevator controls and traffic light controls. In addition to a general purpose computer, this unit could also serve as a general purpose device controller.

Most folks in the industry at the time believed that a single chip computer was something that might be achieved in the "someday" time frame. Ted Hoff thought that "someday" had come. He understood that to build anything on one chip, however, you must be very hardware efficient.

Hoff's background included some heavy experience with the hardware design of the IBM 1620 and the DEC PDP-8 minicomputer, a very hardware-efficient machine, Hoff thought that without much more complexity than a memory chip, a simple processing unit could be built. His knowledge of the PDP-8 greatly influenced his work.

After getting the approval of Busicom to design and build the unit, Hoff and four other engineers including Federico Faggin, whose initials are etched in the first Intel processor chip, completed this project (design and creation) in four months. In 1971, the project was completed. The team, along with Busicom had in fact, created the world's first microprocessor. The end result was the Intel 4004 microprocessor, which goes down in history as Intel's first commercial processor chip. The project was a big success. The future would be even bigger.

Figure 9-1 Enlarged View of Intel 4004 Chip

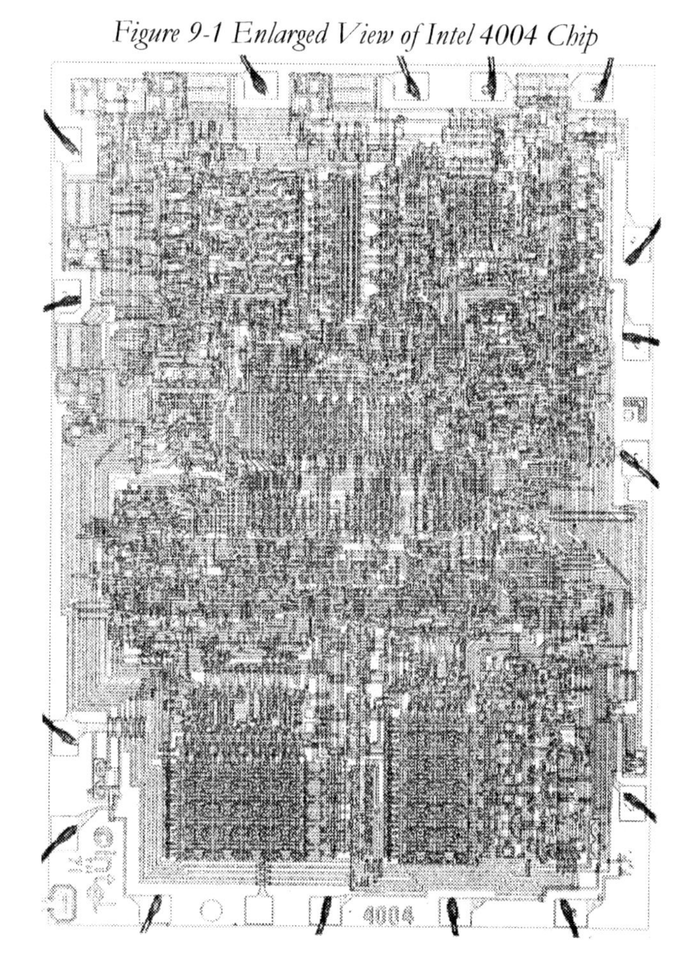

The 4004, shown in Figure 9-1, was not much larger than a fingernail and it was smaller than a thumbnail (3mm x 4mm). On this one chip lived 2300 transistors. Intel sold the chip for $200.00 at the time. This one 4-bit computer chip delivered as much computational power as Eckert and Mauchly's prized ENIAC of the 1940's. However, the 4004 didn't need 18,000 vacuum tubes or the space of a small warehouse to get its job done... just a little less than a thumbprint.

Who Owns the End Product?

Not only was the project a big success, but the R&D was paid for by Busicom. Such a deal! But that created its own issues since Intel was constrained from selling the fruits of its labor. Busicom partially owned the end product. Through negotiations, concessions, and money, Intel was able to get back the rights to its development, and as some may say... the rest is history.

Undisputed Most Successful Chip Maker in the World

Today, Intel is the undisputed most successful personal computer chip maker in the world. They also make lots of other stuff besides processors. They are continually building plant capacity for their semiconductor business as well as other businesses in which they engage. In fact, on July 25, 2005 Intel announced that it is building a new $3 billion plant in Chandler, Arizona and they are refurbishing a plant in Rio Rancho, New Mexico. The company continues to be a major market force to be reckoned with by all.

More Intel Microprocessors All the Time

From the 4004, the company built the 8-bit 8008 (200KHz cycle time) in 1972, which was an 8-bit version of the 4004. Then, in 1974, with the introduction of the 8080, Intel increased the chip density from that of the 8008 by a factor of 2 (5000 transistors) and tweaked the speed to render 20 times more power than the 4004. It did not take long for this full-bodied processor to find itself as the main component of the first kit computer - the Altair. From here, the home computing phenomenon was born.

Intel's Processors Get Even More Power

In 1976, Intel had already begun to refine the architecture of its systems as it introduced the 4.77 MHZ 8085 microprocessor. At about the same time, one of Intel's big competitors back then, Zilog, under the auspices of Federico Faggin, released its famous 2.5-MHz Z80 chip. Regardless of its options, IBM chose Intel chips and became a valued Intel customer. The PC was not the first product for which IBM came to Intel. IBM's first processor decision was the Intel 8085 for its 5280 Intelligent Data Entry Unit, a replacement machine for the popular IBM 3741 Data Station (keyboard to 8" diskette unit).

In 1978, Intel used the 8085, its 8-bit entry as the basis for the introduction of the more powerful 16-bit, 4.77-MHz 8086 microprocessor. When IBM later introduced its DisplayWriter Word Processor, it chose the 8086 as the engine to drive the new unit. This machine had a similar cycle time to the 8085 but, as a 16-bit machine, it worked on twice as many data bits at one time. Thus, in many ways, it was twice as fast as the 8085.

Figure 9-2 IBM DisplayWriter Word Processing System / 8086-based

Then Intel created the hybrid that took over the world. Intel's 8088 was a hybrid of the 8085/8086 unit. In essence it was a cheaper version. Like the 8086, it processed sixteen bits internally. But, like the 8085, it processed 8 bits externally when it went to the 8-bit bus for input/output operations. Within two years, the 8088 chip became the most pervasive of all time, when IBM selected it as the engine for its first Personal Computer (PC).

Figure 9-3 IBM PC with Intel 8088 Chip Circa 1981

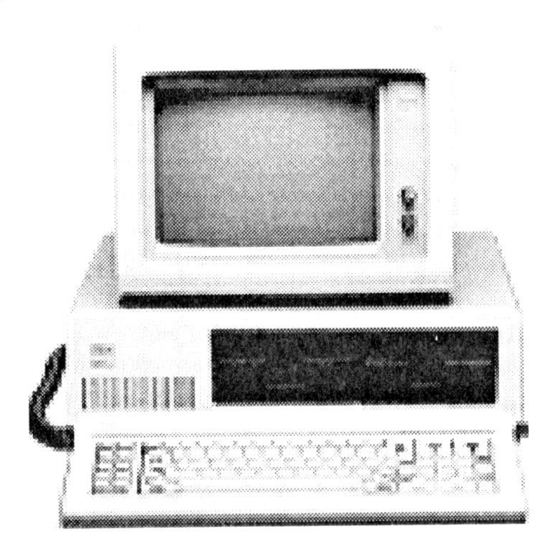

The introduction of the IBM PC in 1981 and its clones in 1982, positioned Intel as a dominant player in the semiconductor industry. One can only speculate where IBM and Intel would be today if IBM had selected one of its own home grown chips, the Z80, or a Motorola chip instead of using the Intel 8088.

Hi Tech Intel iAPX-432 Processor Chip

In 1980, even before its 1978 vintage 8088 was used to launch Intel to a vaunted level of success that even its founders could not have imagined, the chip maker introduced its first 32-bit microprocessor, the iAPX-432. At the same time, it launched a new math coprocessor known as the 8087. Intel hoped that the iAPX 432 would be its beat-all design for the 1980s. It was a very capable design. The company added many expensive facilities to the three chip CPU including advanced multitasking and memory management features in hardware.

In fact, Intel knew they had something special and company personnel referred to this chip as a micromainframe.

The chip set was way, way ahead of its time. Because of all its built-in facility, the processor's data structure support allowed for operating systems to use far less program code than other CPUs of the day. Unfortunately, the design of the chip set was so highly complex compared to the mainstream, Intel's engineers were unable to translate their design into an efficient implementation using the semiconductor technology of the day. The result from all the complex design was a CPU that was too far ahead of semiconductor technology to be implemented. So, as a force fit, it was very slow and expensive. Thus, even before the x86 architecture took off exponentially with PCs everywhere, Intel's first set of plans to replace its x86 architecture with the iAPX 432 ended miserably.

It's hard for me to believe that this happened way back in 1980, just two years after the 8088 and two years after IBM announced its revolutionary System/38. Intel independently tried to create a machine that was so much like the System/38 in its conception that one might have supposed IBM and Intel were collaborating. The iAPX 432, like the IBM System38 (the predecessor to IBM's AS/400, iSeries and i5), was object oriented at its heart and it implemented many operating system functions in hardware and microcode. These functions included process scheduling, inter-process communication, and storage allocation. The integration of such software operations in hardware is particularly impressive especially with the small sophistication level of all other hardware processors of the day.

Since the iAPX 432 had so many features of the IBM System/38, it is a worthwhile comparison. The System/38 in contrast to the 32-bit iAPX 432 was built on 48-bit hardware with 128-bit abstraction. In other words, the System/38 had an abstraction layer that made it think that it was actually working with 128-bits. Considering that today's largest systems are no greater than 64-bits, this is quite impressive.

System/38 programs actually were compiled to believe that the box had a 128-bit address space. Ironically, just like Intel with its 432, in 1978, IBM could not make its System/38 work. Scheduled for 1979 first deliveries, the box was so complex that IBM had to call as many experts as it could find in the company to Rochester Minnesota to make the machine work as intended. I spent a few weeks in Rochester myself during this period, working on migration strategies for IBM's older small business brands. It took another year for IBM to iron out the kinks before System/38's began to land in IBM customer shops.

Lucky Lucky Intel

The iAPX 432 did not have such good fortune. It was scrapped and it has stayed in the boneyard. There are some parallels with the Itanium and the iAPX 432. But, now is not the time in this book to digress. The fact is however, that the combination of the hard work and the hope that Intel had for its iAPX 432 and the fact that the 432 was such a dismal failure could have put the company under. However, luck was on its side. In 1980, IBM chose to use the Intel 8088 microprocessor as the processor for its about to be designed personal computer system.

The fact that IBM, a "real" computer company, would begin building PCs legitimized the market itself. Up to that point in 1981, businesses tended to ignore PCs and mostly treated them as toys with which nerdy engineers liked to play. VisiCalc was the key motivator for accountants to get PCs and this was enough that the market took off in a big way. From then on, there was nothing that stopped Intel.

The iAPX 432 got retired at an early age and it gave Intel its first major marketing failure – but nobody noticed. Now, the 432 lives quietly in retirement along with the Ford Edsel and the Sony BetaMax, and perhaps Elvis. Its unrelated cousin, the IBM System/38 built with similar computer science magic, continues its existence as one of IBM's most popular eServers, rechristened in 2004 as the iSeries i5.

Intel on the Move - More Microprocessors

Intel was not to be stopped. Once IBM got them going in 1981 with tons of 8088 orders, Intel ran and ran and ran... mostly to the bank. They are still running. Intel was already successful when IBM chose the 8088 for its highly successful Personal Computer, but there is no question that the PC was the business event that catapulted the company to the stars. Intel is now just about a $40 billion company.

Just one year after the Intel 8088 was selected for the IBM PC introduction, the Santa Clara chip maker introduced its "almost 32-bit" 80286. This was a more natural growth system for the x86 line than the iAPX 432 and it permitted Intel based older software to run with the new. It also removed some of the hardware limitations of 1MB of total memory addressability, which had begun to plague the 8085/86/88 processor lines.

Computer buffs with some wear on their treads may recall that Microsoft's DOS, the prevalent operating system at the time, had played games with Intel's 1 million available addresses on the 8088. The amount of DOS memory left for users was addressable below memory address 640,000. Microsoft used the addresses above 640,000 to 1,000,000 to get at the memory on adapter cards such as video cards and sound cards and networking cards (Ethernet and Token Ring).

It was not at all straight forward, however, and it made working with the pre-Windows 95 PCs much more difficult than it should have been. And, of course because of the tricks that applications had to play with memory on these DOS / Windows machines, the machines were prone to just stop running and "hang." But, when that happened, Intel always got a pass and Microsoft got the blame.

The 80286 used a larger address and was able to reach 16 MB of "real" addressable memory and 1 GB of virtual memory. Though billed as a 32-bit machine, as the company's 32-bit entrée', the 286 was not Intel's finest hour. However, it was not that bad for a first try and it was much better than the 8088. To put the growth of this processor in perspective, it consisted of 130,000 transistors and it ran at a speed of 12 MHz.

In 1985, Intel introduced the 80386, a real 32-bit computer system with the same power as minicomputers of the day. The company had clearly solved the problems prevalent with the 80286 architecture and had built a fine unit. This processor brought powerful computing to the masses. Again this power was achieved according to Moore's Law, as Intel was able to jam 275,000 transistors on a little piece of silicon. Formally christened the 80386, most users referred to the chip simply as the 386.

In 1989, Intel again raised the bar. The price/performance curve continued its steep climb adhering as usual to Moore's Law and its assumed corollaries. Its new 80486 (486) was substantially faster than the 386 and it featured more than a million transistors, an unprecedented density at the time, for any processor.

Intel Introduces Its Pentium Line

In 1993 as it was obvious that Intel was becoming a large, successful company, other microcomputer manufacturers such as AMD and Cyrix were doing a reasonable job of cloning Intel processors, and selling them for less and less. Cyrix withered away after being acquired by National Semiconductor in 1997 and then sold to Via in 1999, but AMD is still very much alive. AMD had gotten its start as a second source for Intel chips as IBM required, but Intel understandably was not about to give away any of its secrets.

In addition to cloning processors, prior to Pentium, AMD and Cyrix were also cloning Intel's names. These chip cloning companies had begun to use the numbers 486 in their product names so it made it easier for the public to understand there was 486-equivalency in their clone units. For example, AMD had its own 386 called the AM386 and it had a 486 called the AM486. It was a good deal for AMD but not such good brand recognition for Intel.

Rather than introduce an Intel 80586, as expected, to avoid the confusion and resulting lost sales to the clone manufacturers, Intel took the 5 in the middle of 80586 and called its new line the Pentium. Numbers could not be trademarked, but the word Pentium could be trademarked. Intel added system level functions to the Pentium chips, which were above and beyond the capabilities of the 486, and the company promoted a major increase in raw computer power. The Pentium supported multimedia, graphics, fast I/O, and communications facilities that were new to the PC arena. Additionally, while designing the powerful Pentium processor, Intel again followed the laws of their co-founder Gordon Moore, and produced a unit with over 3 million transistors.

Pentium Pro

In 1995, Intel was at it again with the introduction of the Pentium "Pro" processor chip, which featured dynamic instruction execution and other performance-oriented features in addition to more integrated cache memory. Again, Moore's Law was not violated as the Pentium Pro checked in with 5.5 million transistors.

Note: Cache memory is used to speed up disk operations by retaining recently used information in fast memory rather than having to retrieve it from disk for each access. It is a small, fast, expensive memory area that holds recently accessed data, designed to speed up subsequent access to the same data. Cache is typically used between a processor and main memory. There are three levels of cache: primary L1, secondary L2 and tertiary, L3. The lower number cache is always the faster and more expensive, and it is typically closest to the processor. For example L1 cache in modern chips actually runs at CPU speed.

Pentium II

In 1997, Intel introduced yet another Pentium processor. Since it was clearly a new machine and Intel was not about to call the new chip the "Sexium," with Pentium already being a successful brand, this one got the handle, the *Pentium II*. It included a feature which Intel dubbed MMX, which stands for multimedia extensions. This new, beefed up Pentium with its MMX video technology boasted over 7.5 million transistors on a single chip. It also offered improved cache memory enhancements.

Overall, the Pentium II delivered somewhat faster processing than previous Intel units for ordinary processing. But, the MMX add-on made the chip fast enough and powerful enough to support the editing of digital pictures as well as multimedia productions such as music or home movies. For the canine lovers out there, let me apologize for suggesting that prior Pentium models did not have the juice to drive fancy video applications. In other words, for multimedia, the Pentium ran like a "dog."

Pentium III

In early 1999, Intel announced the Pentium III family of high speed processors with 8.5 million transistors and 70 new machine instructions to further assist in the use of the Pentium for multimedia and games processing. The 500 MHz Pentium IIIs that were benchmarked proved to be approximately 10% faster than the 450 MHz Pentium II. Thus, Pentium III by itself at its introduction did not seem to deliver substantially more power than Pentium II processors. However, when applications were written specifically for the new advance instructions buried on the chip, Pentium IIIs ran substantially faster than Pentium IIs.

In October 1999, the unthinkable and perhaps impossible happened. Advanced Micro Devices' (AMD) introduced its line of 700 MHz Athlon chips, which had for the first time eclipsed Pentium chips in performance. It was not long before Intel juiced up the Pentium line again and a game of leapfrog performance leadership began in x86 processor land. In March, 2000 AMD passed Intel again with the introduction of the industry's 1st GHZ processor (1000 MHz). Intel gained the lead again and the game continues to this day.

Intel's Pentium III chips eventually achieved densities of over 28 million transistors using a new .18 micron spacing process. Gordon Moore's law thus continued as the chip law of the land.

Pentium 4

Back when the Pentium 4 (also called the P4) first came out on April 23, 2001, it was received with mixed reviews. The last thing any company needs with a new product introduction is bad press. Yet, that is the only ink that Intel received for some time after the P4 introduction – until they fixed both the problem and the perception.

Clock Speed Hoax

Their new Pentium 4 architecture known as NetBurst was acclaimed when the chip was introduced but after the computer geeks and nerds in the public sector examined it more closely; it was seen as part of the great clock speed hoax of 2001. With a 1.7 GHz clock speed, the new

Pentium 4 did not perform as well as the Pentium III at ½ the clock speed. Thus, Intel suffered from a high degree of well-observed negative publicity. Early benchmarks proved that the Pentium 4's performance was clearly clock-for-clock worse than that of the Pentium III. Even the bump in clock speed to 2.0 GHz in August, 2001 did not save the user community. So, Pentium III processors kept selling for some time thereafter. It was an embarrassing situation for Intel.

As it turns out, according to computer geeks on the Internet, Intel had made some design decisions in the Pentium 4 to produce good clock speed numbers at the expense of good performance. Since machines until then had been evaluated mostly by their clock cycle time measured in thousands of cycles per second (hertz), Intel marketing chose to play to the perception of power rather than provide power.

The marketing folks at Intel prevailed over the technical gurus thinking the technique would go unnoticed by John Q Public. Since PC buyers had related immediately to the clock speed rating in megahertz (1 million processor cycles per second), Intel marketing felt that if the machine had a high clock speed, it would sell better against the competition because the public would not know that the numbers did not mean what they once meant. The Computer geeks claimed in their "blogs" and Q & A forums that Intel marketing had rigged the numbers. As it turns out, the public learned quickly and the Pentium 4 did not do well in sales until Intel addressed performance with models with higher clock speeds and real performance boosts.

Big Densities

Intel's Pentium 4 chips initially achieved densities of over 28 million transistors using the same new .18 micron spacing process as the Pentium III. The chip also contained 16K instruction Level 1 cache as well as 16K data cache. For the non technical, in almost all cases, the more cache the better and there was more cache on the Pentium 4. Putting cache on the chip itself also made the cache perform better. A feature called Hyper-Threading was added to the line in 2002 to help the processor execute two threads at the same time (two different instructions in two different programs at the same time). Again, Gordon Moore's law continued as the chip law of the land.

While Intel was slugging it out to make the Pentium 4 the machine that it should be, the company continued to make certain "models" of its Pentium III line perform better. The Intel Pentium III Xeon processor, originally announced in 1999 extended Intel's chip offerings and was enhanced to provide a needed performance boost for the Pentium III line. The Xeon processor was expensive however, because it used huge amounts of cache memory to achieve its substantial performance boost. The Pentium III Xeon was so good that it outperformed the early Pentium 4 models and was the preferred processor for servers that had major workloads.

Pentium Class Machines - 2001 to 2005

From its 1.7 GHz (1700MHz) initial model, Intel has succeeded in jacking up the clock speed of the Pentium 4 in small increments. The last increment brought the speed to 3.8 GHz in the fall 2004. In Chapter 16, we look much more closely at the issues that Intel faced during this time period in trying to get the Pentium 4 to this higher clock speed.

In the early 2000's, Moore's Law worked fine in terms of Intel being able to add circuits more densely on the processor chip. For example, the latest standard Pentium chips contain about 125 million transistors

In 2004, Intel successfully reverse engineered AMD64 technology and announced its own version called EM64T. This gave x86 applications running on Intel the opportunity for 64 and 32-bit computing on the same chip. With its marketing clout Intel soon became the leader in shipments for 64-bit x86-style computing, a notion originally conceived and engineered by AMD.

In May, 2005, Intel upped the ante again on Moore's Law with an almost doubling of the transistors on its newest "dual core" processors. These behemoth chips checked in at a whopping 230 million transistors. It is getting really hard to believe that all those circuits can live on just one little chip. Intel's new Itanium chip effort is also paying off big in terms of massive densities as the Itanium approaches 1 billion transistors on a single chip. Dual core Itanium models were scheduled to arrive in late 2005.

In its most simple terms, a dual-core CPU or processor chip as announced by Intel in 2005 combines two independent processors (two computers) and their respective caches and cache controllers onto a single silicon die, or integrated circuit – i.e. on one chip.

Intel Dual-Core Processor-Powered PC Systems First to Market

Knowing that AMD was about to launch its dual core processors on April 21, 2005, Intel beat AMD to market. On April 18, 2005, the giant chipmaker announced that computer manufacturers Alienware, Dell and Velocity Micro were authorized to begin selling desktop PCs and workstations based on Intel's first dual-core processor-based platform.

In terms of its formal announcement of dual-core and the availability to ship immediately, Intel trumped AMD's efforts that had long been planned for April 21. In this instance, Intel out-AMD'd AMD.and won the day.

Its first dual-core product shipped was called the Pentium Extreme Edition 840. It was equipped with two processor cores, each running at 3.2GHz. More significant than the 840, the chip king also announced that it would be replacing its Pentium 4 processor in many PCs starting in June 2005 when it introduced its new, mainstream, high volume dual-core 64-bit Pentium D processor.

The Impact of Dual Core

There is no question that the launch of this first dual-core x86 chip is the beginning of a significant transition within the PC / chip industry. Intel has abandoned its philosophy of gaining performance mostly through increases in clock speed. Over the years, Intel was always able to achieve much of its processor performance merely by ramping up the clock speed. Ironically, this dual core announcement occurred on the day before the 40th anniversary of Moore's Law. Intel had used the corollaries to this law for years as circuit sizes decreased to gain the most clock speed from all of its processor designs. Though the law is still alive and well, even permitting multiple cores on one chip, this time the payback is not visible in clock speed

Since the easy performance gains of the past thirty years are gone, Intel is now working hard to assure that it has what it takes on its chips to permit more work to be done with smaller circuits and less dependence on raw clock speed. Newer technologies for Intel such as multi-core, multiprocessing, Hyper-Threading, and other performance innovations are the wave of the future for the giant chipmaker. Along with whatever innovations its engineers bring to the table, these new on-chip parallel processing facilities are destined to be Intel's trademark in executing Moore's Law in the future – certainly over the next several years.

> Note: See parallelism considerations in Chapter 20 for a look at what this change may mean to Intel users and developers..

Intel Ready to Roll

At its dual core chip announcement, Don MacDonald, Intel's vice president of its Digital Home Group noted that "this milestone has additional significance for Intel given that it's the eve of the 40th anniversary of Moore's Law. With dual-core processor-based system availability, today is a historic date for the computer industry as PCs begin having 'two brains' instead of one. Moore's Law is about giving you more computing capabilities without an equivalent increase in cost. This is what bringing dual-core processors to PCs is all about."

In backing up its rhetoric with concrete plans, in April 2005 for example, Intel had no less than 15 new processor chips in development. All are based on dual-core or multi-core technology. The future looks very exciting for Intel. Considering all the pain that Intel absorbed to assure a launch of its dual core technology just days before the AMD launch, you can bet that the Chip Wars are still on and battle plans are burning hot in both companies.

If this book were intended solely for computer geeks and nerds, I would take the time to explain all fifteen plus planned processors, provide the vital statistics of each and offer various forms of commentary on their intended markets. That would be very boring and since I have more fun things in store for us as we explore Intel's current battle with AMD as well as the quiet battle looming in the background with IBM. Besides, the computer geeks and nerds can get all the information they want about Intel's plans on Intel's Web site – www.intel.com

Other Mainline Intel Processors - Xeon

In 2001, Intel came out with its Pentium IV class Xeon processor chip. The company dropped the Pentium part from the name and it now calls this processor the Intel Xeon. It is targeted for high-performance and mid-range, dual-processor workstations. Intel'a dual and multi-processor server configurations of Xeons appeared later in 2005. Intel Xeon processor-based workstations achieve performance increases between 30 and 90 percent over Intel Pentium III Xeon processors, (depending on applications and configurations.) Just as the Pentium 4, the processor is based on a new Intel innovation that the company calls its NetBurst architecture.

The Intel Celeron Line for the Budget Minded

Shortly before the Pentium III processor was announced, in what the trust-busters might call a smart anti-competitive move, in June 1998, Intel introduced a new processor model called the Celeron. Celeron chips at the time were based on the same architecture as the Pentium II microprocessor, and they continue to be designed for low-cost PCs. They run at somewhat lower clock speeds (266 and 300 MHz for example back then) and at the time of their announcement, they were not as expandable as Pentium II microprocessors.

The initial Celeron processor did not do too well because it was actually very slow. Celeron was not supposed to have the bells and whistles of the Pentium but the initial chip actually ran so poorly that it did not sell. Intel had based the chip on its Pentium II but it ripped off all of the L2 cache that had made the P2 run so well. Since a good cache implementation often gives as much of a performance boost to

PCs as a high clock speed, the no cache Celeron model ran like an old dog.

Intel had introduced the Celeron to crush the clone manufacturers - AMD and Cyrix. AMD is still out there fighting but the Celeron and AMD basically did Cyrix in. Intel put the Celeron together to get business from the PC users who would not pay Intel's tariff for a Pentium model, but wanted a Pentium class machine. In other words, it was built for PC buyers who might gravitate to the AMD or Cyrix chips for an overall lower price rather than pay for a Pentium.

By caching data in fast, pricey memory, you speed up your system performance. A secondary cache is bigger than the primary cache (usually in the same chip as the CPU) and fits between it and main memory (RAM). It's faster than main memory, but slower than primary cache memory. Celeron cache has typically been around 256K in size, though early Celerons were not quite so generous with L2 cache.

Other than the missing cache, a problem that was corrected in the next Celeron iteration, it has been really tough to find much of a difference between the Celeron processors and the Pentiums. However, if you're from Intel, Celeron's make big sense. It was a brilliant marketing move. Intel disabled a few circuits on the Pentium and got a Celeron. For the most part, users who could not afford a Pentium class machine had little problem getting "Intel Inside" with the Celeron. For Intel that saved a lot of chip sales from going to AMD or Cyrix.

Celerons have come a long way since Intel first introduced the units. The original units had zero-cache, and though very inexpensive, you got what you paid for. Intel has released many new Celeron models over the years and they also have added just enough L2 cache to the mix now to make it an even better deal.

Recognizing the value of the brand name, Intel has kept the Celeron processor name in tact with every major Pentium processor change. The company merely re-issued the Celeron with new enhancements to parallel the corresponding Pentium of the time. In 2000, for example, the Celeron II was released. It was basically a lower-cost version of the Pentium III, just as the first Celeron was a stripped down Pentium II. Celeron II eventually was able to use the Pentium III true 100 MHz bus when Intel increased its clock speed to 800MHz at the beginning of 2001. In 2002, Intel again re-issued the Celeron using the Pentium IV as a basis. The chip again contained 128KB of L2 cache. So it is a far cry form the doggy days of the late 1990's.

It did not take long for Intel to realize the brilliance of its Celeron marketing move. By mid 2000, Intel's Celeron chip line, just two years old, had become the number 2 chip family in the world. Today, the Celeron continues right behind the Intel Pentium 4 in sales. Celerons often cost as little as 1/3 of the price of Pentiums so it is easy to see that the chip has accomplished its mission of gaining back much of the ground Intel had been losing to the chip clone vendors

HP & Intel Rolling For 64-bits

In the very early 1990's Hewlett Packard had a great idea to begin using a leading edge technology known simply as Very Long Instruction Word or VLIW in a new brand of processors. Unlike CISC or RISC, VLIW is its own way of computing and it is a notion espoused for supercomputing over the years. Instead of a variable instruction word as in CISC or a short constant instruction word as in RISC, VLIW, as designed by HP was to use a long, constant size instruction word and place a number of instructions within the word. The overall idea of VLIW is to get the most operations as possible operating in parallel in one core. To pull off a technology such as this, of course, many other enhancements had to be made to the chip. HP knew this was a great idea but also knew that it would take substantial resources to pull it off so HP approached Intel and the two became partners in a new chip building endeavor.

Many readers may recall through experience or through prior reading that IBM was not doing too well as a company in the early 1990s. Some even go so far as to say that IBM's own pride did it in. IBM had often thought it was the best and felt it needed little to no help to achieve its objectives. As the financial roof fell on the company, humility also set in big time. Big Blue had already farmed out the processor of the PC to Intel and Intel was doing very well. IBM's first attempt to capitalize on RISC technology with its RT/PC in the mid to late 1980's had already failed miserably as the chip performed like a dog. Other than its mainframe line, IBM seemed to have lost its way in being able to build competitive processors that could stand well against Intel, Sun, HP, and others. Cash strapped as it was, IBM was quite vulnerable.

The only hope for IBM to survive in the mid-range processor business was for the company to design a chip that would beat its competition dead. As the RT/PC was announced and it did so poorly, IBM began an internal effort called Project America to produce what it hoped would be the fastest RISC processor ever created. The processor became known as the PowerPC. A more complete story of this project is given in Chapter 11.

In the early 1990's, IBM was willing to do lots of unusual things to assure its survival. Though the initial results of the America project in the form of the RS/60000 (used the PowerPC chip) system were in and IBM had already convinced Apple and Motorola to form the AIM alliance for PowerPC, IBM had little confidence in itself. The PowerPC project was clearly a success as the RS/6000 was starting to shatter some performance records, but it had cost IBM lots of cash to create the processor family, and John Akers, IBM's Chairman at the time was not looking to keep spending money.

So, soon after IBM was calling on Apple to get Jobs and company to use its PowerPC chips, HP and Intel, armed with their new VLIW design, called on IBM to convince IBM's executives to use their soon to be available VLIW processor chip. Basically, the duo offered that IBM would no longer have to make processors for any of its hardware products, including the mainframe, AS/400, and RS/6000, because the new VLIW processor was designed to be earth-shattering in function and performance. Their message was that no effort on IBM's part would be able to overcome the lead that Intel and HP felt they had in technology and they almost had IBM's management set to sign on the dotted line. That's how good Intel's new box looked as designed. IBM was just about prepared to drop its own chip development and move to the latest and greatest that HP and Intel expected to have on-line in just a few years.

To get the right feeling for this point in time, all this was happening when IBM was at its worst financially and the company was preparing to dump its chairman while the chairman was trying to raise cash by selling off parts of the company. To make matters worse, though the Internet and dot com fever running wild, IBM systems were not participating. IBM stuck to its proprietary communications technology. IBM had chosen not to spend the development dollars to adjust its AS/400 and its mainframe lines for the Internet. Big Blue was lost and its product line offered little hope. Sun dominated the Internet field and the Intel / HP Consortium was threatening to take over IBM's server market as soon as its new chip was available.

In 1994, as IBM was preparing to introduce its 64-bit RISC-based AS/400 and RS/6000 lines of computers, the fate of IBM as a producer of its own chips hung in the balance. Intel had recognized that its longstanding x86 single thread uniprocessor design was not going to carry the day so, as noted previously, in a historic move, the company joined with HP in a cooperative agreement to develop a common 64-bit VLIW microprocessor architecture to be introduced soon after the turn of the century. The two chip designs that came from this effort were the Merced (Itanium), and the second stage McKinley (Itanium II) chips. Many industry analysts and a number of IBM executives believed that these HP/Intel innovations had a shot at defining computing in the 21st century. For IBM it must have been a chilling thought.

Intel Itanium Processor

The Itanium processor design appeared to be at a first class level to IBM. However, when Lou Gerstner took over IBM in 2003, he thought for himself. Eventually, he was convinced by some IBM engineers not to bet the IBM Company's hardware future on a product built by another company that was not yet ready.

In 2001, Intel brought out its initial iteration of the long awaited Itanium processor as the first in a family of 64-bit VLIW based products. Itanium was designed as the replacement for all Pentiums, both servers and desktops. Unfortunately for Intel, the 2001 edition of Itanium was not ready for prime time. It had many technical issues and there was little software that would run on the box, and so it created little marketing hype and was basically a dud in most respects.

2002: Itanium 2 Processor

In 2002, Intel released the Itanium 2 processor as the second member of the Itanium processor family, a line of enterprise-class processors. This processor corrected many of the issues in the first iteration. Intel built some great performance into the machine as in 2005, an Itanium powered supercomputer was in the top 10 supercomputers in the world. But, Intel has had a difficult time convincing its customers to sell the product.

When a long-expected product comes to market and it does not measure up immediately, such as the first generation Itanium, it is very difficult for any company to overcome the negative inertia caused by what some might call a failure. Along the way, all the wonderful plusses of having the benefit of the doubt disappear once a product does not measure up to expectations. Potential customers can easily become cynics rather than supporters. When the first Itanium sunk, it was a devastating marketing happening for Intel. Competitors jeered and called the machine the "Itanic." Now that the second Itanium stands well on its own, few Intel customers seem to care, and that is even worse. The future of Itanium is to be decided.

The Intel Future...

Though Moore's Law has done Intel well over the course of the company's history, the company did begin to prepare for the day that "more" Moore would not be gained from simply ramping up clock speed. The biggest of all preparations, as noted above, came in the 1990s when Intel partnered with Hewlett Packard to build the next generation 64-bit VLIW processor.

With Itanium having its trouble in the marketplace, HP dropped out of the alliance in 2004. Moreover, with EM64T as a competitor to its own Itanium, Intel is now the world's foremost supplier of 64/32-bit x86 chips. By anybody's calculations, this is a marketing triumph. EM64T technology is now at the heart of all of the company's new chip efforts. With a competing 64-bit technology that is selling well, it will be difficult for Itanium to find its way into the mainstream product line.

For the 2000 - 2010 decade, Intel expects to be shipping machines like the Pentium 4 type with multiple cores. The first big batch of these should be under the Pentium D umbrella and Intel expects that this will enable the company to continue to leverage corporate Windows and Linux application software investments.

Will DOS and 8088 programs survive?

One of the hallmarks of Intel technology over the years, other than the Itanium, has been that it has clung on to the past. Theoretically a DOS program from the 8088 genre of computing could execute on a standard Pentium 4 machine with little issue. The 64-bit x86 version of Windows XP, which runs the EM64T extensions, is another firm step away from the DOS computing legacy. In much the same way as Windows 95 weaned even multimedia and game fans from the command prompt to the graphical user interface, EM64T and the ability to run 32-bit and 64-bit applications on the same Windows machines precludes the ability to run old DOS 16-bit applications in any mode. DOS is gone. DOS is dead except on old technology.

Just about all current 32-bit software runs fine under Win XP for 64-Bit Extended Systems, but older 16-bit and DOS-based applications are left out in the cold. In this day and age, this really is not such a huge deal. Those looking to run ancient DOS apps are certainly not the target market for a spanking new Pentium D system. Of course, if you need 16-bit capability, the message is to make sure you get yourself some old spank-less technology in-house before you can't get it anymore.

Of course, 16-bit and 32-bit standard x86 computing is not even a possibility on the Itanium chip. So, Moore's Law and the law of diminishing returns (profits) on old technology have de-motivated the technology suppliers from providing DOS on future machines. To say the same thing in circles, there will be no speeding tickets issued to DOS applications for using Moore's Law's corollaries to get the best out of the new technology. DOS won't even be allowed on the highway. The message of course is: "Say good-by to DOS!"

What's Next for Intel?

In 2005 and beyond, Intel has changed its computing paradigm and is beginning to focus on parallelism. There's a lot more about parallelism and the Chip Wars with AMD as you continue with this book. Intel processors will be dual core first and then multi-core.

Intel will achieve aggregate performance boosts for applications that can benefit from parallel processing. Since single thread high clock speed processing had been Intel's trademark until recently, the future may be somewhat uncertain for Intel for a short time, while the market decides how it can best use its new multi-core parallel processor technology. But, overall, you can bet that Intel will survive and prosper.

Chapter 10 The Challenger: Advanced Micro Devices (AMD)

Where Did AMD Come From?

About five years ago I spent several months researching the origins of computing for a book I was writing at the time. I did not go as far back as Francis Bacon and I did not include the abacus in my studies but I most certainly examined the mainframe manufacturers (the BUNCH), the minicomputer manufacturers, and the PC manufacturers. In order to do a complete job, I traced technology all the way back to the creation of the transistor at Bell Labs (Shockley) and went on to the advent of semiconductors and the major players in the semiconductor industry.

I have been in the industry myself since 1969 when I began my career with IBM so I am not really a neophyte to the players who have been such a key part of this industry. I never finished the book but as I look back I am somewhat embarrassed that I completely missed AMD as a major player with similar historical roots to the billionaire computer icons who now grace the technology landscape.

Ironically, the company I missed, AMD is a major player today in the chip wars but even more than that, the company's roots go back as far as Intel's at Fairchild. Moreover, the company's founder Jerry Sanders and others who helped him get AMD going were part of the esteemed graduates of Fairchild Semiconductor, the company, which "spawned" the likes of Gordon Moore, Robert Noyce, and Andy Grove.

For about fifteen years now, Advanced Micro Devices (AMD) has produced "x86" processors as a rival to those of Intel. From its beginning as a second source (licensed production) manufacturer of processors for Intel, AMD has developed its own powerful range of processors of independent design. In fact, recently AMD designs have gotten so good that to remain compatible with Windows, Intel was forced to copy it. Amazingly, in order for Intel to keep up with AMD's technology, the company was compelled by industry pressure to clone the AMD way.

When you research AMD, you are immediately impressed by the company's strong commitment to people. Having worked for IBM, a company known for treating employees exceptionally well during the Watson "entrepreneurial" years, it is refreshing to see that a people-first company can still make it in the modern industrial world. After employees, the next item on AMD's list of priorities is its customers. AMD has so much information about its attempts to please its customers on its Web site that at first it strikes one as merely self-serving rhetoric – apparently disingenuous. Then, it becomes easy to see that right after employees comes the company's respect for its customers and the many things (many are mutually beneficial) that AMD does to assure that it pleases its customers and satisfies their needs.

Jerry Sanders Is Quite a Guy

AMD history is marked by a commitment to innovation that's truly useful for customers. AMD principal founder Jerry Sanders, a very impressive person in his own right, has always maintained that "customers should come first, at every stage of a company's activities." The current CEO, Hector de Jesus Ruiz, carries the same torch, saying, "Customer-centric innovation is the pre-eminent value at AMD. It is our reason for being and our strategy for success."

On May 1, 1969, Jerry Sanders and seven friends founded Advanced Micro Devices in the living room of one of the co-founders. Though there were seven founders of AMD, Sanders was the guy who everybody felt could raise the money to get the company moving. Sanders had been a sales engineer with the gift of gab. He raised the money. He wanted to be President. He became President. Then, before the cash started to arrive the seven co-founders thought they would give back some stock options to the investors to get more needed cash with which to run the business.

Sanders thought otherwise. They outvoted Sanders seven to one on how to proceed but the brash 31-year old slammed back at them that they had no vote and that the only vote that counted was 1 for and 0 against. Jerry Sanders again found a way to get what he wanted without having to give anything up. From that point on, Sanders ran the company. Unlike Moore and Noyce and Grove at Intel most people who write about AMD, state that Jerry Sanders and seven others founded the company. The others are rarely mentioned by name. That's how powerful a manager Jerry Sanders was at AMD.

AMD History

Before the end of 1971, AMD's first product ran off its Fab1 facility. It was the AM9300, a 4-bit MSI shift register. AMD was off and running. In 1972 Jerry Sanders took the company public.

To help AMD customers stay ahead of the innovation curve, Sanders and company launched its first memory product in 1975 — a RAM (random-access memory) chip known as the Am9102. In the same year, the company began to sell a processor with a familiar name, the 8080A standard processor. AMD reverse-engineered Intel's 8080A standard processor but they had a little help being Intel's required second source. IBM would not do business with Intel unless there was a second source.

This product boosted AMD into the microprocessor arena and it provided a competitive alternative to the market. Despite the recession in 1974-75, AMD's business grew to $168 million, representing an average annual compound growth rate of over 60 percent.

By the end of fiscal year 1981, sales had more than doubled over 1979 figures. To help meet customer demand, AMD expanded plants and facilities worldwide, particularly in Texas, with new production facilities in San Antonio and additional fabrication space in Austin.

Quality Standard

Technology innovations were rocketing along at an unprecedented rate, and AMD was dedicating resources to help its customers keep up. To that end, the company made a significant investment in research and development, and introduced a notion called "INT.STD.1000" — the industry's highest manufacturing quality standard at the time. In 1984, AMD celebrated its 15th anniversary with one of the best sales performances in company history. By 1984, the company was also listed as one of the top 100 companies to work for in America

AMD Doing Well

In 1985 AMD made it into the Fortune 500 and in 1987 it merged with Monolithic Memories to create an even more powerful company. In 1988, AMD began its Submicron Development Center, which would become the driving force in the company's process-technology innovation efforts throughout the '90s. "Process technology" refers to the multitude of technologies used before, during, and after the process of manufacturing a product. AMD believes that its process-technology leadership has enabled the company to consistently provide its customers with valuable competitive advantages.

Early Intel Battle

The late '80s and early '90s marked a significant turning point for AMD. During eight years of legal wrangling with Intel, the company struggled to uphold cross-licensing agreements that enabled AMD (and other companies) to compete and provide competitive alternatives in the PC semiconductor market. Using such approaches, AMD believes it broke the monopoly on 386 chips. AMD won the lawsuit.

The Name Game

Along the way to being very successful, Jerry Sanders had a number of very creative, yet very simple ideas. For example, there was much copying in the technology sector back then. For example, if TI would come out with a part called the TI6539, Motorola might create the same part but call it the MC2203. This confused customers who sometimes could not find the proper cross-reference information to be able to purchase the same parts from multiple vendors. Jerry Sanders did not play by the same rules. If TI had a part called the TI6539, and AMD chose to make the same part, using Jerry's nomenclature, the part would be the AM6539. Though AMD was sued by a number of companies over this bold decision, AMD prevailed since the part was distinguished by the AM prefix meaning AMD.

AMD Is Really Catching Up

In 1991 AMD began head to head competition with Intel in the 386 space with the introduction of the AM386 Microprocessor family. In addition to just offering a product, AMD continued to offer higher clock speeds and lower power dissipation than the competition. AMD shipped more than one million units of the AM386 processor family by the end of the year 1992. In 1993, the company launched the first members of the next-generation AM486 processor family, which went on to power Compaq computers and thousands of others. And when AMD technically won the legal battle against Intel, it made it all worthwhile and helped establish AMD as a for-real player in the industry. You can imagine Intel's frustration in having its own 486 machine be copied by AMD and also to have AMD use the same numbers to identify its version. That's of course one of the reasons why the Intel 586 chip was named the Pentium rather than the 586.

Figure 10-1 AMD K6 Pentium II Clone

To deliver the true innovations that Jerry Sanders believed his customers deserved, AMD began forging critical relationships with infrastructure, software, technology, and OEM partners — the cornerstone of what AMD called its "connected business model." First, AMD partnered with NextGen to create the AMD-K6 microprocessor family, designed to restore competition to the supply chain and offer an outstanding solution for desktop computing. AMD later acquired NextGen. After its introduction in 1997, the return of competition against Intel helped drive costs low enough to enable a whole new market segment — PCs that were affordable for everyday people. Sanders aim was to offer the highest performance and lowest cost solution for the masses. Now, AMD and IBM have a major partnership that is paying off big time for AMD.

IBM In the Background

AMD and IBM have had agreements in the past and the most notable is the one signed in 2003, which ultimately resulted in the Opteron and some technology breakthroughs. IBM is a very secretive company and overall, the company keeps good secrets. However, there was a major leak in IBM in the 1993 time period. Around this time, IBM was trying to recover from having been made irrelevant in the PC industry by its then partners, Intel and Microsoft. Big Blue was trying to figure out how to not lose the desktop completely as terminals were being replaced by PCs left and tight.

Among a number of notions, some engineers at IBM believed that they could reverse engineer the x86 instruction set and that they could build a chip that ran x86 instructions better than Intel's own processors. So the rumor mill has it that in 1993, developers at IBM's Burlington, Vermont facility started to work on a version of the PowerPC that would place the x86 instruction set on silicon along with the PowerPC set. It was actually a brilliant idea but at the time IBM literally had no money.

The Intel x86 instructions would be built directly into the CPU, not as an adjunct. While the work was done by IBM, independent of Apple or Motorola, (See Chapter 15), in other words, without the support of the AIM alliance, this chip began to be known inside IBM and by the media as the PowerPC 615. However, cash concerns and early performance issues in the switching between the x86 and native PowerPC instruction sets resulted in the project being canceled in 1995 after only a limited number of chips were produced for in-house testing. A second version of the chip was designed but never fabricated when IBM canceled the project.

IBMers report that they have earrings and other jewelry with these already diced PowerPC 615 chips as the major ornament. The chip could be both a PowerPC and a PC. Its only problem was cleaning out the instruction pipeline as it switched from one mode (x86) to PowerPC or vice versa. With the density achievements in today's processors; and with Linux, this act would now be quite achievable. Back then, as the story goes, Microsoft was not too willing to run a version of NT in the PowerPC mode and a Windows operating system on the x86 side of the chip at the same time. It would be necessary for a dynamic OS switch. In 1995, there was no notion of hypervisor control in microprocessors so this would have been a Microsoft OS task..

The rest of the possibilities are purely speculative. IBM is so secretive; I could get neither denial nor agreement to this speculation. As you are about to read, AMD got real good at cloning Intel's chips, including its x86 instruction set. There is a possibility that IBM's work on the 601 found a home at AMD and that has helped the second largest PC chipmaker in its recent successes against Intel. Is it coincidental that IBM, the leader in 64-bit computing shows up and AMD comes out with a 64-bit chip that also just happens to have the 32-bit x86 instruction set etched in silicon. Then again, it is just speculation.

Viva la Clone

In the late 1990's AMD was on a roll. It acquired NextGen, a "clone" chip manufacturer in 1996 and then in 1997, the company introduced its first Pentium Clone, the AMD-K6. The technology for the K6 came from NextGen. The AMD-K6-2 processor featured something called 3DNow! Technology. Invented by AMD, 3DNow! technology was the first x86 innovation to significantly enhance 3D graphics, multimedia, and other floating-point-intensive applications for Windows-compatible PCs. AMD continues to make improvements to the x86 instruction set even today. The AMD-K6-2 processor offered a tremendous competitive difference to customers developing leading-edge hardware and software, and it helped lay the groundwork for the future of AMD and other processor designs.

AMD Overtakes Intel

In 1999, AMD raised the ante in its battle with Intel as it introduced its Athlon processor line of "seventh generation" x86 processor chips. This was the culmination of a company dream — to design and produce an industry-leading, proprietary Microsoft Windows-compatible processor. AMD notes that the AMD Athlon processor empowered its customers with a compelling new alternative, the first-ever solution to use chipsets and motherboards optimized specifically for AMD's processor. The AMD Athlon won many awards and it really knocked Intel on its socks. It proved itself to be an outstanding platform across a wide variety of high-performance applications.

AMD Achieves the Impossible – First 1GHz Chip

In January, 2000, Dr. Hector de Jesus Ruiz became the President and Chief Operating Officer of AMD on his way to bigger and better things. Jerry Sanders was 63 and was preparing for retirement. Ruiz was his successor designate. In March 2000, AMD shocked the world again as the company that beat Intel to the magic 1GHz level in performance with its Athlon line. That's one billion clock cycles per second. Jerry Sanders of AMD called that "the PC Industry's equivalent of breaking the sound barrier."

Additionally, AMD made great strides in its memory business with Double Data Rate memory (DDR) This increased peak data throughput to the processor by up to 100% at comparable costs. This enabled AMD customers to boost performance tremendously without sacrificing their bottom line.

AMD was also successful in its Flash Memory business. The company introduced highly advanced Flash memory products — the 32-megabit Am29BDS323 and the 64-megabit Am29BDS643 — which enabled AMD customers to create next-generation cellular-phone applications featuring Internet connectivity, PDA functionality, video streaming, and even MP3 capability.

More Plants Coming

Quite notably, AMD's ongoing relationship with Fujitsu saw the companies break ground on a third chip fab. In retrospect, Sanders notes that "in hindsight, much of 2000 was "the calm before the storm" — which would come barreling down on AMD customers in 2001." Luckily for AMD, some of the changes that were already underway helped place the company in a position of strength and agility to handle the tough years about to come.

MirrorBit Technology

From 2001 to 2003, during the worldwide economic slowdown, AMD revenue dropped substantially. In 2001, innovating through its adversity, AMD introduced its first multiprocessor chip, the Athlon MP processor along with a number of significant new memory products. Early in 2001, AMD made a major announcement – a breakthrough in memory-cell architecture known as MirrorBit, which offered impressive benefits. MirrorBit is a proprietary AMD architecture that enables a Flash memory product to hold twice as much data as standard Flash memory, without compromising overall performance. With AMD's new MirrorBit technology-based solutions, companies could add features to their cellular phones, PDAs, and a host of other products, in a cost-effective manner.

Industry Goes Bust

You may recall how bad things got. The bust that actually began in 2000 turned into a major break in 2001, hitting the "dot-com" companies first, and eventually spreading to AMD's customers in areas such as networking, computer manufacturing, telecom, and business in general. Even the glimmers of hope seemed to disappear after the horrific events of September 11, as political uncertainty deepened the industry decline. The semiconductor industry overall experienced the sharpest drop in history, as customers struggled to find new ways to remain competitive without buying new technology.

Sanders was always the eternal optimist and he knew that despite the challenges, new opportunities were emerging, if one knew where to look. Sanders charge to the troops was: "We must out-innovate the competition, delivering semiconductor solutions based on better ideas that increase the performance, reduce the cost, and shorten the time to market for our customers' products."

Hector Ruiz CEO

In April 2002 Hector Ruiz succeeded Jerry Sanders as CEO. 2002 was a watershed year for AMD, as many of the company's long-standing philosophies and the current market conditions synergized into opportunity. As Hector Ruiz assumed the role of CEO, he coined the term "customer-centric innovation." This term basically reduced into a simple phrase what AMD always stood for

On September 23, 2002, Hector Ruiz announced AMD's "connected business model" philosophy to the general public, saying "To be successful, we believe that semiconductor companies must build relationships with customers and partners that truly blur the lines ... where the companies are connected and invested in each other's success."

As a real pick-me-upper in 2002, AMD won a prestigious contract to provide the processing power for a Cray supercomputing solution, which supports the National Nuclear Security Administration's Advanced Simulation and Computing program, or ASCI, which helps ensure the safety and reliability of the nation's nuclear weapons stockpile.

In January, 2003, AMD and IBM began a formal partnership. The two signed a groundbreaking joint technology agreement. In April 2003, AMD introduced its phenomenally successful Opteron processor and its innovative AMD64 technology and architecture. This landmark made AMD the first x86 chip vendor to support both 32-bit and 64-bit computing on the same chip with no user programming changes.

In September, 2003, AMD beefed up its Athlon processor with the addition of the AMD64 technology making the Athlon a 64-bit contender in the x86 workstation world. In November 2003, recognizing that it had become a winner in the chip business and hoping to get more of Intel's share, AMD held a groundbreaking for a new chip fab in Dresden Germany

Overall, in 2003, the technology industry regained its footing, as stronger consumer demand and a more confident global economy drove tech stocks higher. AMD benefited with an up-tick in its gross sales volumes. PC sales grew strongly, wireless computing gained momentum, and 64-bit computing entered the mainstream.

AMD was on a real collaboration theme for 2003. Hector Ruiz set the tone stressing AMD's belief that early and open collaboration with partners and customers is the critical driver of true innovation. The

AMD partnership with IBM to jointly develop advanced chip technologies is extremely significant.

Figure 10-2 The Inimitable Jerry Sanders

Sanders Steps Down

In 2004 AMD announced that Jerry Sanders was stepping down as AMD's chairman. Sanders was given the honorary title of "Chairman Emeritus," but gave up all his power and authority. Hector Ruiz, as expected, became chairman. He began to hold all the reins at AMD. Sanders noted that he had hand-picked Ruiz. So, now Hector Ruiz is the Chairman, Chief Executive Officer, and President of AMD. At the meeting for the changing of the guard, Jerry Sanders got the opportunity to introduce Hector Ruiz. He noted that Ruiz "is ideally suited to take AMD to the next level."

In his hey day, Sanders was known for wearing flashy suits and making memorable quotes. He was at his best again at this meeting. In past speeches, for example, Sanders had called the Pentium 4 "a dud," described the Pentium 4 core as "friggin huge," and once asked a reporter, "How do you spell bullsh*t?" As always, Jerry Sanders was many things: charming, caustic, stubborn, bristly, enigmatic, compelling, colorful, determined, and ambitious. Yet it was the calm, sedate Ruiz who had orchestrated the successful launch of AMD64 and the new Opteron/Athlon 64-bit processors. It was Ruiz who returned AMD to profitability. Sanders is now able to enjoy a comfortable retirement. As one would expect for the founder, AMD continues to pay him handsome compensation for his time with the company. And Jerry Sanders is still watching.

Besides the changing of the guard, AMD had some major milestones in 2004. The company saw its AMD64 architecture reach the pinnacle of success. By February, more than 1,000 OEMs, software and hardware developers responded to customer demand for AMD64 processors. Among these was the Lenova group who, perhaps anticipating its own efforts to buy the IBM PC Division, introduced the AMD Athlon™ 64 and AMD Athlon XP processor-based "Lenovo Feng Xing V" series of home personal computers. These boxes are distributed in China.

AMD also introduced its Sempron Processor Family and the Sempron processors, a new family of value processors that are intended by AMD to redefine everyday computing for value-conscious buyers of desktop and notebook PCs.

In the Dual Core race, in mid 2004, anticipating its April 2005 release, AMD demonstrated the industry's first x86 dual-core processor for both 32- and 64-bit computing. The chips were housed in an HP ProLiant DL585 server powered by four dual-core AMD Opteron processors.

Figure 10--3 AMD Opteron Dual-Core Design

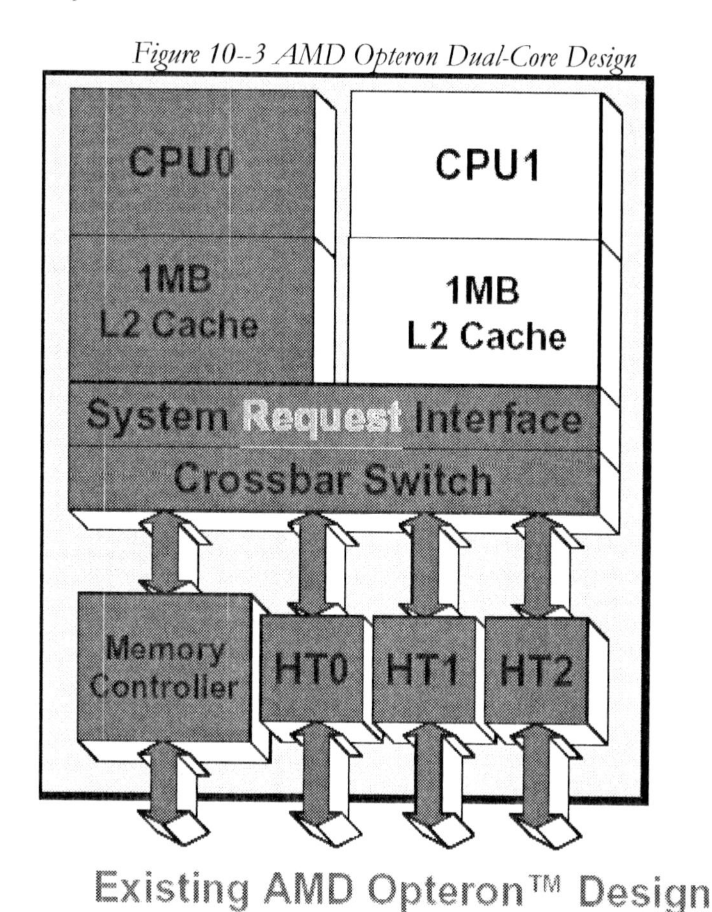

As noted previously, AMD is one of the world's leading suppliers of flash memory. In 2002, the company announced a new technology for flash that it called MirrorBit. In November, 2004, AMD unveiled its plans to move MirrorBit way out in the future with a notion the company calls Spansion Vision. This is AMD's vision to scale its MirrorBit technology to 8 gigabit densities on 65-nanometer lithography by 2007.

AMD began 2005 with major accolades. High-tech market research firm In-Stat selected the AMD Athlon 64 FX-55 processor as best desktop processor of 2004. In-Stat Microprocessor Report recognized the AMD Athlon 64 FX processor for its superior performance, feature-rich architecture, and support for 32- and 64-bit capability and enhanced virus protection. For a company that once was viewed only as a "clone" maker, AMD had come a long way. Reaching the $5 billion mark in sales for 2004 was another crowning achievement announced in January 2005.

AMD and Microsoft

In February 2005, AMD took on a role that would typically have been reserved for Intel as the company partnered with Microsoft to move enterprise customers to 64-bit computing. Thus, it was an AMD Opteron processor that powered the first x86-based 64-bit servers in the Microsoft Technology Centers.

Microsoft used HP ProLiant DL145 and DL585 servers powered by the AMD Opteron processor at its MTC locations in North America, Europe and Asia. These systems enable MTC customers to migrate, test and validate 64-bit applications while preserving their investment in 32-bit applications. Microsoft praised the AMD Opteron in these two public statements: (1) "The impressive combination of the AMD Opteron processor and Microsoft's 64-bit operating systems brings the flexibility and scalability needed for enterprises to realize the full benefits of 64-bit computing." (2) "Using AMD Opteron processor-powered servers in Microsoft Technology Centers was an obvious choice since the centers were created to bring the best hardware, software and expertise together to solve customers' problems. With AMD64 technology in the MTC, our teams will have the ability to help optimize customers' 32- and 64-bit .NET applications all on a single platform."

Dual-Core Surprise

On April 21, 2005 AMD announced that it was introducing the world's first 64-Bit, x86 multi-core processors (dual core in this case) for servers and workstations while celebrating the second-anniversary of its 64-bit AMD Opteron processor.

Unfortunately for AMD, Intel's PR people were working double-time and just three days earlier, on Monday, April 18, Intel formally introduced its dual core processors, trumping the AMD announcement. Since AMD had been shipping test versions of its dual core offerings before Intel and the company was more prepared than Intel to enter the market on April 21st, we offer that both Intel and AMD concurrently announced dual core x86 processors in April, 2005. It is safe to say that if Intel had not reverse engineered AMD64 technology, it would not have done so well in matching AMD's dual core efforts.

With its April 21, 2005 announcement, AMD again showed its technology leadership in the x86 industry by providing the performance and power benefits of multi-core processors. Migrating to AMD's dual-core processing based on AMD64 technology can be quite seamless, thanks to its non-disruptive upgrade path since the chip was designed to deliver the higher performance of a multi-core processor in the same power envelope and the same infrastructure as a single-core AMD64 processor. For AMD's server/workstation customers this means that the AMD Opteron processor can continue to provide the best performance-per-watt ratio in the industry. For a company about 1/8 the size of the industry sales-leader Intel, having such a technologically advanced solution in a market defined by Intel is substantial.

AMD has partnered with Microsoft to assure that its designs are fully supported in the pervasive Windows server marketplace. Without Microsoft's cooperation, the 64-bit and the dual core Opteron would have the same fate as other chip designs that are incompatible with Windows workstation and server operating systems. With the availability of Windows Server 2003 x64 editions and Windows XP Professional 64-bit x 86 Edition (both upgraded to support AMD64 and its dual core in 2005), customers now have a mainstream, industry-standard 64-bit operating system optimized to take advantage of the performance and efficiency improvements of AMD64 dual-core technology. The simultaneous availability of AMD64 dual-core processors and Windows 64/32-bit edition represents a huge performance leap and much greater value for Microsoft customers for the future. In fact, Microsoft offered its sincere congratulations to AMD on its leadership in delivering such innovative technologies to the market.

Dual-Core on the Desktop

Raising personal computing to new heights, in June 2005, AMD also upgraded its Athlon 64 X2 processor chip brand creating a dual-core desktop chip as well as a desktop-replacement notebook processor.

One of the major benefits of dual core AMD64 processing on the desktop is that it permits many software applications to run simultaneously. Additionally it enables consumers and businesses to take digital content creation and multimedia experiences to a whole new level. PC users, who have often been frustrated by staring at the Windows hourglass icon when trying to work on multiple programs at once, can significantly increase performance with the AMD Athlon 64 X2 Dual-Core processor. For example, PC users can simultaneously burn a CD, check e-mail, edit a digital photo and run virus protection – all without the system seeming to fall asleep in the background.

AMD Has Come a Long Way

From a company best know for imitating other company's products to a company whose inner beliefs have spawned phenomenal innovation, AMD has certainly come a long way. From a bit player continually on the verge of financial collapse to an architecture leader in an industry created by its biggest competitor, AMD has not only accepted the technology challenge, it has exceeded everybody's expectations.

AMD is not only excelling in processor technology but the company is also firing on all cylinders in the memory market with its superior MirrorBit technology. However, as important as memory is to AMD's future success, its legacy now is that the company has forced its major competitor Intel, in order to keep up in an industry, which Intel itself created and dominated, to clone the clone. If imitation is the finest form of flattery, then no longer having to imitate but to be imitated is the finest form of innovation. Kudos to AMD.

Chapter 11 The Silent Contender: International Business Machines (IBM) Corporation

The IBM Company and its Beginning

To the third millennium and beyond, IBM continues to be a great company with great products and great opportunities. The company traces its "computing" roots back to Dr. Herman Hollerith and the invention of the 80 column punched card. It was Herman Hollerith's inventions, which led to the processing of the 1890 census in 2.5 years using electro-mechanical machines. The census had been taking 10 years to complete before this improvement. Hollerith, a statistician for the Census Bureau formed the Tabulating Machine Company, which later combined with the Computing Scale Co. of America and the International Time Recording Co. to form the C-T-R Company.

Thomas Watson Sr. Takes the Reins

In 1914, Thomas Watson Sr. at age 40, was fired from his job as general sales manager of the National Cash Register Co., after a dispute with the president, John H. Patterson. He joined C-T-R as general manager. Watson adopted some of Patterson's more effective business tactics to help move the combined business forward. These included: generous sales incentives, an insistence on well-groomed, dark-suited salesmen and an evangelical fervor for instilling company pride and loyalty in every worker. He preached a positive outlook, and his favorite slogan, "THINK," became the marching orders for C-T-R's employees.

Watson also stressed the importance of the customer. He well understood that the success of the customer translated into the success of the company. He treated employees with the highest regard, and in return, he received their full loyalty and hard work.

He became President within 11 months of joining the company. In 1917 in Canada, and 1924 in the U.S., Watson changed C-T-R's name to the International Business Machines Company (IBM) to better reflect his business expansion goals for the company.

IBM Did the Right Thing for Employees During the Depression

During the Great Depression of the 1930s, IBM kept its workers employed by building new machines even though the demand was slack. Meanwhile, of course, the rest of the economy was floundering. Undaunted by the hard economic times, Watson continued to take care of his employees. Watson's IBM was also noted for having been a real leader in giving benefits to employees. IBM always took pride in being a leader in providing a secure position with an accent on the family. For example, in the mid 1930's the company was a leader in providing life insurance, survivor's benefits, and even paid vacations.

IBM's altruism to its employees during the depression not only paid off in employee morale but also prepared IBM for what was to come. The Social Security Act of 1935 brought with it the requirement for substantial data processing. Watson grabbed a landmark contract with the government to maintain the employment records for 26 million workers. IBM was able to fulfill the contract with the inventory the company had built during the lean years. The lean years were over. Orders from other U.S. Government departments and businesses continued to make the company a success.

The Early IBM Product Line

The products in the IBM's line at the time were big, and bulky electromechanical 80 column tabulating (TAB) card machines. These consisted of keypunches, sorters, collators, reproducers, interpreters, calculators, and accounting machines. IBM thrived on equipment that today can only be found in museums. Figures 11-1 and 11-2 show pictures of two of these behemoth electromechanical machines – an IBM 082 Sorter and an IBM 407 Accounting Machine.

Figure 11-1 IBM 082 Sorter

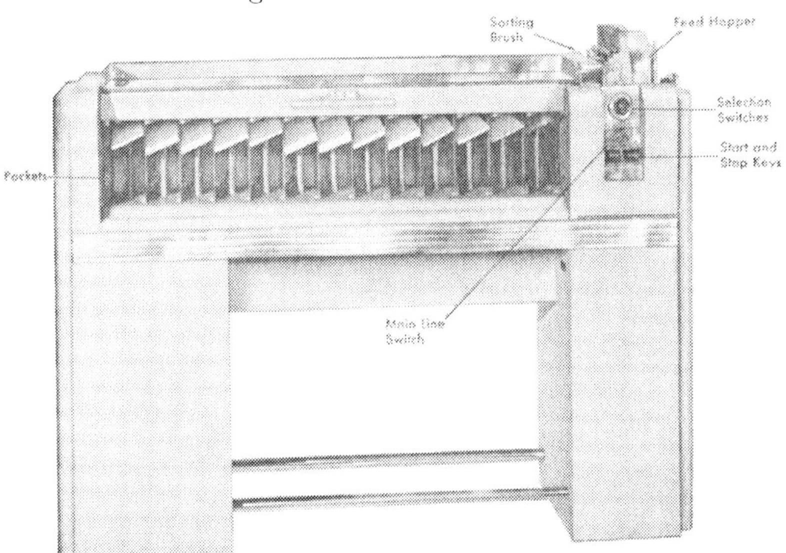

IBM continued to rent this TAB equipment to its customers well into the 1970s. It was extremely profitable. Much of the equipment had been in service for many years and had paid for itself many times over. Each additional day was found money for IBM. Those were surely the good old days for IBM.

Figure 11-2 IBM 407 Accounting Machine

IBM Did the Right Thing for Employees during World War II

When World War II began, Watson volunteered all IBM facilities for the use of the U.S. government. IBM plants were used to make bombsights, rifles, engine parts and several dozen other hardware items necessary for the war effort. The same Thomas Watson Sr., who pioneered IBM's favorable employee benefits plan, used the same sense of kindness and grace to establish a fund for widows and orphans of IBM war casualties, using the nominal one-percent profit on those war products as a means to finance this benevolence.

Automatic Sequence Controlled Calculator

During the war years, IBM still was able to engage in R&D. The work of Harvard's Howard Aiken, working with IBM during this period produced a one of a kind computer prototype called the IBM Automatic Sequence Controlled Calculator (ASCC), which was also called the Harvard Mark I. It was completed in 1944 after six years of development with Harvard University. Aiken's work is considered to be the first digital computer, though its architecture differs significantly from modern machines.

Thomas J. Watson Jr. Assumes the Presidency and the Chair

In the 1940's there were a number of worldwide efforts to create a computer using the Von Neuman concept of stored programs. The most notable of these projects were the ENIAC and EDVAC efforts of John Mauchly and J. Presper Eckert. The ENIAC and the EDVAC were recognized as two of the first computers ever built using the leading edge Von Neuman concept.

Mauchly and Eckert later brought their skills to Remington Rand, IBM's biggest competitor at the time. IBM did not pick up the torch for building computers until just before Thomas Watson Jr. became the President of the Company in 1952. Watson Jr. was expected to become chairman and he eventually assumed the CEO spot in 1956. But, in the late 1940's when the major groundwork for computers was underway, without Watson Jr. in control of IBM, the accolades for the first commercially available computer went to Remington Rand not IBM, as the company introduced the first commercially viable computer known as the Univac I. in 1951.

IBM's First Mass-Produced Computer

In 1952, IBM introduced the IBM 701 as its first large commercial computer. It was based on the vacuum tube. The mainframe era was just getting started. This machine used much smaller tubes and was much more nimble than IBM's Mark I. But this machine would not capture any markets for Big Blue. The machine that took the industry by storm and became the most popular computer of its day was the IBM 650. It was smaller and even more agile than the 701. It was introduced in December 1953. By 1956 over five-hundred 650s had been produced, clearly establishing the IBM 650 as the industry sales leader. Though many more than a thousand computers are sold in an hour today, at this stage of computer history, 500 was a huge number of units for any company to produce in a several year time period. The success of the IBM 650 signaled the world that the computer revolution had begun.

Business Expansion Overseas

Arthur K. Watson, Tom's younger brother began the World Trade Corporation. This IBM division handled the company's foreign operations. Though not as well known as his older brother, Arthur Watson's achievements catapulted IBM into a dominant position in Europe. For example, in the 1960's IBM installed over 90% of all computers in Europe.

Transistor Technology

Those of us, who were around then, may remember the tiny radios in the late 1950s. Most of these seemed to be imported from Japan. You may recall these were called "transistor" radios. They were the rage. In 1959, IBM began to take advantage of this technology in computers instead of using the bulkier, less reliable and more costly vacuum tubes of the day. Transistors performed the same switching functions as tubes, but were substantially smaller and permitted processing at far greater speeds. In 1959, the IBM Company introduced its 7000 series of mainframes which were the company's first transistorized computers.

The most powerful of the 7000 series was the 7030 which was also known as the "Stretch." Some of the other notable IBM units of this transistor era were the 1401 business computer and the 1620 scientific computer. The 1401 became the most popular business mainframe computer of the day, and the 1620 took off in the scientific arena.

The technology advances had some drawbacks, however. Each generation of computer hardware required a totally new set of programs (software). To move from the IBM 650 (very popular vacuum tube model), to the transistor built IBM 1401 for example, all programs had to be redone. IBM was looking for a way to build a family of computers which performed scientific operations and business operations equally well.

IBM Announces the System/360

Its efforts paid off on April 7, 1964, when IBM was able to introduce the System/360, the first large "family" of mainframe computers to use interchangeable software and peripheral equipment. The word System was selected because the whole product line was devised as a system, in which all models and all parts are interrelated. The number 360 was chosen because of the 360 degrees in a circle. The system thus had an all-encompassing objective. All applications, even those with diverse purposes, such as those with a business and/or scientific aspect, were to be accomplishable using the System/360.

Besides providing substantial growth without reprogramming when moving throughout the family, System/360 designers accommodated the notion of upgrading computers rather than scrapping the old and purchasing the new. With System/360, IBM customers could simply upgrade parts of their hardware. Moreover, the company developed microprogramming based emulation facilities to enable older IBM programs built for other machines such as the 1401 to be able to perform well in the new hardware architecture. It was so all encompassing that IBM almost did not get it out the door. Fortune magazine dubbed it "IBM's $5 billion gamble." It was a gamble, however, which fortunately for IBM paid off in a big way.

IBM leveraged its System/360 gamble even further in 1970, with the introduction of the System/370 mainframe family of computers. The company continually enhanced this mainframe line of computers, upgrading the family every five to seven years. Along the way, the company abandoned the word *system* in front of its mainframe models and began to use four digit numbers instead, such as 3090, as product family names rather system names as in System/370. In the mid 1990's IBM returned to those more simple days as the company announced a newer, more powerful set of computers, which it called the System/390 (S/390). The System/390s were pegged as IBM's mainframes for the millennium.

Once the year 2000 had come however, IBM changed the mainframe names again as well as all the names of all its computer servers. On

October 3, 2000, the company introduced the eServer zSeries of computers, as a re-branding of its System/390 product line.

In the mainframe hardware and software world, IBM is by far the champion. Its recent competition has been Hitachi and Amdahl (Fujitsu), and Unisys. In early 2000, due to strong technical innovations and major price reductions, Hitachi gave up. They were unable to compete with IBM in the mainframe space, and they discontinued their mainframe line of computers. That left IBM's only mainframe competition as Amdahl, which is wholly owned by Fujitsu. In 2001, Fujitsu packed it in and chose not to compete in the 64-bit mainframe world. Though Unisys likes to point out that it is a traditional mainframe house, the company's most recent efforts have been with Intel based microprocessors.

As other forms of microcomputer based processors have achieved mainframe speeds, IBM remains the one company offering a traditional mainframe approach using custom designed mainframe hardware and operating system software. IBM's clear dominance in this market forced all others to pack up their bags and head for the hills. One thing is for sure: "IBM knows mainframes."

In 2005 IBM announced its newest mainframe the z9, which continues to be based on a traditional mainframe processor chip. The industry expects that in the next several years, IBM will abandon its very expensive mainframe-only chip technology in favor of the next generation of IBM's Power Architecture. (See Chapter 15.) The current generation POWER5 processor running in IBM's iSeries had been benchmarking twice as fast as the fastest mainframe but with the new z9, it checks in at about the same computing performance. Mainframe performance is expected to benefit if and when the company converts its mainframe operating systems to the POWER line. Moreover, the company will save a tremendous amount on R&D for its mainframe unique chip, while maintaining its performance edge.

IBM Announces its Small Business Computers

In 1969, IBM announced the batch computer with the funny little 96-column card. They called it the System/3. This machine was intended for small businesses. As the businesses using System/3 grew in size, they demanded that IBM make the System/3 bigger to help them expand in a non disruptive fashion. IBM was very accommodating.

But there was demand by IBM marketers for an even smaller computer to address the needs of smaller businesses who could not afford a machine with the price tag of a System/3. In 1975, IBM introduced the desk-sized (not desktop) System/32 which was a single station unit with an attached printer.

Figure 11-3 IBM System/32 Desk-Sized Computer Circa 1975

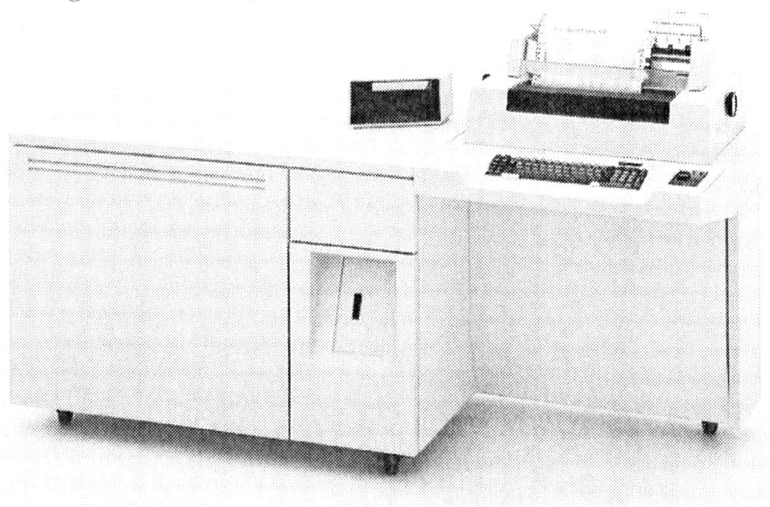

Using the System/32 operating system as a base, in 1977, IBM stole some hardware from its not-yet-announced System/38 and created a multi-user version of its System/32. IBM named this machine the System/34. For its day, System/34 offered revolutionary capability to painlessly support a multi-user terminal environment. Programmers claimed that "you could not help like the System/34."

Figure 11-4 IBM System/34 Multi-Station System Circa 1977

In 1978, IBM announced the System/38. For its day, the System/38 was well ahead of its time. In addition to its 48-bit hardware, the 1978-system had an abstraction layer that presented software with an image that the machine was actually a 128-bit unit. Because of this abstraction, when IBM changed System/38 hardware to the AS/400 and from CISC to RISC and from 48-bits to 64-bits, the customer programs written and compiled for the System/38 continued to run. Today's eServer i5 is a direct descendent of the System/38 and just like its predecessor it offers numerous productivity advantages over all of IBM's other systems. These include an object based architecture, single level storage, an integrated relational database as well as an integrated transaction processing.

In 1983, IBM upgraded the System/34 as the System/36, and added some well needed processing power, disk capacity and additional software facilities such as a full office and word processing package. Many believe that it was the S/36 that legitimized IBM's presence in the midrange marketplace and paved the way for the 1988 introduction of the AS/400 as its follow-on to the System/38 and

System/36. As I recall the sales numbers, at the time the AS/400 was announced, IBM had sold about 20,000 System/38's in the US and an equal number throughout the rest of the world. In contrast, over 200,000 System/36's had been sold worldwide. That level of customer acceptance on the part of the System/36 users provided fertile ground for the early marketing efforts for the AS/400.

Thomas Watson Jr. Steps Down as Chairman and CEO

In 1971, just two years after the introduction of the System/3, Thomas J. Watson Jr. stepped down as Chairman and CEO and was succeeded by T. Vincent Learson, a long-time Watson supporter and loyal IBM executive. At this time, IBM informally began the tradition of CEOs leaving the chair at age 60. Learson, who was 58 at the time of his leadership assumption, turned the reins over to Frank T. Cary in 1973.

Tom Watson Jr. continued his public and business life after IBM retirement. He was appointed U.S. ambassador to the Soviet Union. He served there from 1979 to 1981. He remained a member of IBM's board of directors until he resigned from the post in 1984. On April 1, 1993, Watson accompanied IBM's first outside CEO, Lou Gerstner, to his first day of work.

Thomas Watson Jr. died in late 1993 at the age of 79. On January 5, 1984, IBM memorialized Watson with a full paged advertisement in the Wall Street Journal (page A11). The text of this ad, under a distinguished picture of the former IBM Chairman, captures the essence of the man:

Figure 11-5 Thomas J Watson Wall Street Journal Memorial Text

Thomas Watson Jr.
1914 - 1993

For all his achievements—
as visionary, entrepreneur, corporate leader,
and distinguished statesman—
we will remember Thomas J. Watson Jr. most
for his adventurous spirit,
his innate sense of fair play, and the vigor of his friendship.
We mourn his passing, but
we will forever be grateful that he lived.

Figure 11-6 Thomas J Watson – Wall Street Journal Memorial Picture

When Thomas J. Watson Jr. stepped down for good in 1984, the era of Watson leadership and Watson guidance had come to an end for IBM. Like Jerry Sanders, Robert Noyce, Bill Gates, Michael Dell, Scott McNealy, Larry Ellison, and the other billionaire technology entrepreneurs, Thomas J. Watson Jr. ran IBM like as if he owned the company. That's why his days as the head of IBM were very successful. There are many former IBMers and many others who believe that IBM was at its best during the Watson years. From Watson's perspective everybody was important, from customers to employees, to the families of employees. There was a major difference in IBM from the day Thomas J. Watson Jr. retired.

Competition Made Itself Known

T. Vincent Learson was not around for very long to be noticed by IBM history. However, Frank Cary's was around quite awhile and his tenure was filled with lots of adventure. IBM found itself in a number of new marketplaces, and there was no shortage of excellent ideas and projects within IBM. New competitors seemed to continually emerge from the shadows.

In the very late 1960s and early 1970's, a number of startup companies had emerged as competitors of IBM, but in a completely different way. They called the computers they made minicomputers, but for little machines, they were capable of doing quite a bit of work. Unlike the original BUNCH (Burroughs, Univac, NCR, Control Data, and Honeywell), who competed directly with IBM in the mainframe area, this group did not have large computers in their bag of tricks, so they typically called on small companies and departments of large companies. To make their sales, they used the same successful techniques, which PC vendors, ten years later, would use to get around the MIS hierarchy.

These companies included Data General, Digital Equipment, Hewlett Packard, Prime, and Wang. They built strong product lines and became formidable foes of IBM. Though these minicomputers operated best in a scientific environment, and real time, such as in the control of traffic signals, they were morphed enough to become reasonably capable business machines. During this time, Bell Labs perfected the original Unix operating system, and eventually most of the minicomputer vendors began to offer their systems with a Unix option.

IBM's Minicomputer Hits the Market

In the 1970s and 1980's, IBM used its System/3 and later the System/34, /36,/ 38, and AS/400 product lines to compete for commercial business against the minicomputer competition. In the real-time computing arena, IBM's product was called the System/7, but it was a mainframe type machine, and not viewed by the industry as a real minicomputer.

In the mid-late 1970s, IBM took notice that the bona-fide minicomputer marketplace was doing quite well without its lead or presence. In a knee-jerk reaction, the company rushed to market a phenomenally rich hardware architecture which was completely open and extensible. The machine was called the Series/1. At the time of its announcement, however, the company had not taken the time to create any software for their new unit. There was not even an operating system. Moreover, IBM did not support Unix on the machine. Thus, as one might suspect, the Series/1 initially floundered. Years later after IBM built a few operating systems for the machine and some software vendors wrote some applications for it, the machine began to sell. However, it never achieved the level of success as IBM's competition, especially Digital Equipment Corporation.

Series/1 Exits, Stage Left

The Series/1 was replaced by a second generation RISC machine called the RISC System/6000 or simply the RS/6000. The invention of RISC is one of the biggest secrets in the annals of IBM research. The story of the invention of RISC is given in Chapter 15.

The RS/6000 (pSeries) family was designed to provide a broad range of platforms for engineering/scientific, technical, and multi-user applications. The processors were noted as being especially appropriate for numeric-intensive use.

The RS/6000, now called the pSeries continues as a prominent member of IBM's server line today. It formed the basis for a very special machine (Deep Blue) which, in 1997 gained world recognition in a chess match with Garry Kasparov, the world's best human chess player.

Chairman Opel Ushers In the IBM PC

Frank Cary served until John R. Opel's appointment as CEO in 1981. Opel was caught up in IBM's tremendous mainframe success at the time and believed if manufacturing constraints were lifted, IBM would move from a $50 billion company to a $100 billion company by 1990. Under Opel, IBM made a lot of hard hats happy. The company was continually building new plants and adding on to existing facilities to help make the Chairman's dream of $100 billion revenue by 1990 a reality.

John Opel's appointment coincided with the introduction of the IBM PC. With the Personal Computer, or PC, for the first time, IBM began to enter homes, small businesses and schools. The PC machine had been designed, developed and assembled in little more than a year mostly from available industry piece parts. The only really noticeable IBM part was the Raleigh keyboard. Yet the new home unit was a real computer... and it was an "IBM". It offered 16 kilobytes of memory (expandable to 256 kilobytes), one or two floppy disk drives and an optional green or color monitor. Who needs a hard drive?

Built From Piece Parts

To get these piece parts, the late Philip, D. (Don) Estridge, who was the point man for the IBM effort, along with a group of 12, contracted the production of its components to outside companies. The diskette drives came from Tandon. The keyboard and main circuit board and the little logo came from IBM. The 8088 processor chip came from Intel, and the operating system, called DOS (Disk Operating System), came from a 32-person company called Microsoft, from the absolute other end of the country.

Here Comes the Clones!

It did not take long for the engineers of the electronics world to reengineer the IBM PC. With IBM's premium prices, there was a lot of room for less expensive offerings. Look-alike clones began to appear as early as 1982 and IBM struggled trying to devise clever ways to fight this unwelcome invasion. The first major wave of defense came too late with the 1987 introduction of the PS/2. The PS/2 came with three major technical innovations:

1. Microchannel bus
2. Operating System/2 (OS/2)
3. 3.5" diskette.

Over the years, IBM PCs were consistently outperformed and out-priced by the clone crowd. IBM suffered a drain in its customer set, which began to choose the faster, less expensive clones over IBM's assurances of reliability, availability and good service. To answer the charge, IBM introduced a number of new PC lines which were better and more price competitive. These include the ValuePoint line, IBM's own clone company Ambra, and the IBM Aptiva and NetVista products of just a few years ago..

In the end, IBM could not survive in the industry that it created – personal computers. In early 2005, IBM announced that it was exiting the PC business and it sold what was left of the workstation PC business to Lenova, a Chinese firm.

Another Changing of the Guard - John F. Akers Arrives

John F. Akers took over for John Opel as CEO in 1985. By the time Akers took office, the cash was all promised, and the drawing board demonstrated that IBM was going to have to do a little scrapping to keep itself afloat. Under Opel, the company had decided that it was going to be capacity driven, rather than be like the old IBM which was always supply constrained. However, by building plant capacity without a corresponding sales increase, the company found itself in a strange position – cash constrained.

IBM was not coming close to achieving its share of a booming IT marketplace. Ex-Chairman Opel's $100 billion dream was short lived. Unfortunately for John Akers, as soon as new facilities were ready to be brought on-line, they had to be shut down and/or sold for lack of need and for cash. Since IBM had also ramped up its employee head-count to provide the workforce for the anticipated expansion, this too created stress on the corporation's resources. The new employees were simply not needed.

Problems and Homemade Recipes

John Akers inherited the beginnings of a mess. Since IBM had always had enough cash to end-run any problem in the short-term, there was no time-tested corporate cookbook for working through such a big problem. History shows that Mr. Akers' homemade recipes did not serve the company well. Under Akers, IBM focused on streamlining operations and redeploying resources... such as personnel. A number of back to the field moves were put in place to take excess plant personnel, retrain them in sales, and place them in the field sales organization. The intention was to both give them a job (a vestige of the Watsonian ethic) and to help increase sales. But, to redeploy the masses as IBM did, you need to have plenty of cash. IBM quickly got plenty short of this primary resource.

To keep the company looking like it was doing well while it was really not, John Akers was willing to sell off division by division to supply the company with cash. Whereas John Opel had built the farm, John Akers took IBM awful close to buying the farm.

A Big Bright Spot: The AS/400

When the Application System/400 came out in 1988, this box soon began to make its impact. No other midrange commercial competitive unit was safe against this unique IBM-built powerhouse. Before long, the revenue from AS/400 sales eclipsed that of Digital Equipment Corporation, which, at the time was the second largest computer company in the world. The industry labeled the AS/400 machine the DEC Killer because it almost always won the day in head to head competition. It was IBM's finest machine and the perfect machine with which to run a business.

The success of the AS/400 was responsible to some degree in forestalling John Akers' departure for a few more years. The failures of the small mainframe systems, the long time to market for a real RISC box, and the collapse of IBM as a serious player in the PC marketplace however, offset the AS/400's opportunity from fully rescuing IBM's CEO..

IBM Sells Its Office Products Line

The company had previously spent so much money on plant, equipment, training and redeploying, that when Akers took the helm, as noted above, the cry was for real cash. At first glance, an ideal candidate for the corporate bail-out was discovered. It was IBM's typewriter, keyboard, and small printer business -- the division that created the popular "Selectric" typewriter with its floating "golf ball" type element. With the advent of the PC, IBM management felt this division was no longer strategic to the company's mission. It was deemed expendable and could be sold.

After the deal, the investment firm of Clayton, Dubilier & Rice Inc. walked away with what was to become an independent company. It was dubbed Lexmark Inc., and was and is headquartered in Lexington Kentucky, the former site of IBM's typewriter / small printer plant. In hindsight, it is too bad that IBM had to let Lexmark go. Last time I checked, the company was doing pretty well for itself with about $6 billion in annual sales.

IBM Research Success

Good people are hard to keep down. While Akers was the commander, IBM's research fellows and the company's significant investment in R&D produced four Nobel Prize winners in physics. These were achieved with breakthroughs in mathematics, memory storage, and telecommunications. Of course this work helped IBM make great strides in extending computing capabilities. However, there were no Noble Prize winners in marketing or in strategic planning. Despite IBM's huge technical achievements, the company was having a difficult time breaking even. IBM's research endeavors are covered in more detail in Chapter 24.

IBM Looks to Survive!

IBM was answering its business challenges internally by reducing expenses, primarily by cutting its workforce. The company decided that it could not afford a direct sales force any longer and turned its business over to distributors. Since the local support team was eliminated, the company also reduced expenses by no longer offering free customer support for its long-term customers. While this saved money in the short-term, in the long term, IBM never regained its pre-eminence in customer accounts.

By 1993, Akers' company could not be concerned with what was right for the long term, and the company was frittering away beneath him. He saw survival as IBM's major issue. With annual net losses reaching a record $8 billion by the end of 1993, cost management and streamlining continued as a business necessity. During this period, IBM was examining itself as a large company and had some misgivings about its ability to compete in all areas of the information technology industry. The company began to restructure even more of its major divisions, such as large printers and even storage. The large printer division, for example, was spun off into a separate unit, and IBM appeared ready to further split the company into separate businesses to be sold off.

Itanium Powering IBM Servers?

Since IBM stuck to its proprietary SNA instead of embracing TCP/IP and the Internet, by the time the Internet was in full bloom, it was running mostly on equipment from Sun. IBM had missed the boat. The RT/PC had failed and the Power Architecture, though promising had lots of competition from Sun, the Compaq / Digital Alpha, HP, and of course Intel and its promise of Itanium. IBM management seriously considered giving up on the PowerPC and signing up for the future Itanium.

IBM came very close to abandoning its own chips. It was another business that John Akers could exit and thus reduce expenses. Moreover, though PowerPC was a measured success, the Digital Alpha outperformed it and Intel was preparing to take over the world with its Itanium. For IBM the competition's lead seemed insurmountable at the time. The idea was that all IBM servers would be based on Itanium, which was formally announced in 1994 and was projected for delivery in four years.

IBM eventually listened to its own Gurus who noted that while Intel was the leader in volume, its processor chips are basically intended to power workstations, not multi-user systems. Intel processors had been basic single-thread processors without much support if any for parallel activity. Therefore the IBM Gurus were able to convince IBM management that there was actually a bigger opportunity in continuing its PowerPC architecture than abandoning it for the Itanium. IBM engineers believed that their chips, designed for large servers, would be capable of delivering the power that applications in the next few decades would require. They did not see the Intel Itanium as a choice that would help IBM achieve its server goals.

They convinced IBM's top management to approve the enormous investment (literally billions of dollars) required to originally create and then stay the course with the 64-bit POWER line of processors for use in IBM's major league non-mainframe servers. IBM made a number of risky decisions and by 1998, Big Blue was doing multi-threading with these chips, and by 2001, the company had integrated two CPU cores on every chip. Then in 2004, IBM upped the ante substantially with its POWER5 offering with two central processing units, full simultaneous multithreading, and a reduced pipeline for faster execution of instructions.

IBM's PowerPC a Big Success

IBM did all this while gaining up to four times the performance using a power conserving reasonably low clock speed. What's more, IBM built it so that any existing application on three different operating systems could run on the same processor chip - AIX, Linux and OS/400. In fact, on IBM's iSeries with micropartitioning, all of the operating systems could actually run at the same time on one POWER5. All could take advantage of these tremendous capabilities in POWER5 without changing a single line of code. IBM got it right and with POWER5, the company now has the fastest single chip processor in the world.

Hindsight is wonderful and POWER is one of the good stories but very little seemed like it would survive John Akers' ax. Akers' IBM did not want to be in all of the businesses in which it was engaged because it actually did not know how to run them. When the business report card for Akers would come in year after year IBM was getting deeper in the whole. Many aspects of IBM's business were not holding their own. Akers began to sell assets to bring in cash to meet payroll. He also began preparations to sell additional parts of what had once been the company's core businesses. He was desperate and given more time, IBM would surely have fallen with him.

The fairest thing I can say about Mr. Akers is that he never did recover from the debt left him by his predecessor. In 1985, when Akers, who was 50 years-old at the time, succeeded John Opel as the Chairman and CEO, most IBMers believed that he represented long-term stability for the next ten years until he reached retirement age at sixty years old.

However, by 1993, it was obvious to the world, the industry press, and the IBM Board of Directors that the old methods in IBM would not keep the company in business. Akers stepped down from his responsibilities before it was his natural time. He had placed the company in such a precarious position that, for the first time in its over 80-year history, the IBM Company looked outside for a person with the smarts and the fortitude to bail it from the hopeless quicksand in which it seemed to be immersed.

Lou Gerstner Takes Over

On April 1, 1993, after a few misfires, the IBM Board found such a leader. Louis V. Gerstner Jr., became IBM's chairman and CEO on that day. Thomas Watson Jr. drove him to work that day. Though it was April Fool's Day, Lou Gerstner was no fool. Gerstner's executive experience included eleven years as a top executive at American Express and four years as the CEO of RJR Nabisco during its major restructuring efforts.

Gerstner seemed like an ideal candidate from a business perspective, though some felt his lack of experience in the computer field would limit his ability to take charge and become effective. This proved not to be the case. As a consumer products executive, he already had a customer-oriented sensitivity and he had strong executive and strategic thinking strengths. Lou Gerstner had built these strengths and sharpened them even before his American Express days as a management consultant for McKinsey and Co.

IBM stockholders around the world watched closely. It was not long after he arrived that Gerstner began to take dramatic, unprecedented action to stabilize the company. These steps included rebuilding IBM's product line, while continuing to shrink the workforce and making significant cost reductions. For IBM to survive and thrive, Gerstner believed he had to break the unwritten job security contract and long-standing full-employment practice with his employees, and IBM executed massive layoffs for the first time in its history.

Unlike Akers who had some problems managing the divergent units within IBM, and who was well poised to begin a break-up of the company, Gerstner had a different idea. Though there was mounting pressure, due to IBM's poor record, to split Big Blue into separate, independent companies, Gerstner decided to keep IBM in tact. He believed that one of IBM's biggest strengths was its ability to create and deliver integrated solutions for customers, rather than what could be called the piece-parts solutions du jour, which were served by necessity by IBM's competitors. He decided to keep IBM's unique advantage and move the full company in a direction to suit the times.

When Gerstner came to IBM in early 1993, the business he came to, in many ways was dying. It had been sapped of its financial strength losing $5 billion in 1992 and on its way to a record loss of $8 billion in 1993. It was on its last legs. It was not the tradition-rich IBM, which every other Chairman had inherited. It was not the old IBM. It was a beaten, broken company when Tom Watson Jr. accompanied Lou Gerstner to work on his first day.

Neither Gerstner, nor any other magician, could create the old IBM. It was not his heritage. If there were still an old IBM at the time that Gerstner came aboard, he would not have been the right pick for CEO. The IBM, which Gerstner got, did not have the luxury of a recession-proof rental business. It did not have a crackerjack sales force. It did not have a leading Office division. In fact, the IBM Gerstner got had been prepared itself to be sectioned off, with parts being dissolved or sold for cash

Gerstner inherited an IBM with a death wish marketing strategy on its way to the bone yard. He was the right guy to rescue IBM. He came with no constraints. Though cuts continued, even the IBMers did not have the same disdain for Gerstner's cuts as they did for Akers'. Akers had been one of the team himself and he did not protect the company or the employees. Lou Gerstner was an outsider, brought in to breathe life into a dying company. Employees expected that Lou Gerstner would continue to cut costs. It was his mission.

Gerstner Saves IBM from Bankruptcy

When the cuts were mostly done and IBM began to return to some sense of normalcy, Lou Gerstner had actually saved the company. As much as the prior chairman was determined to split the company into a bunch of tiny and weak IBM's, which could be sold to raise cash, Gerstner was determined to join it back together to form a strong, formidable IBM. After the cuts began to mend as the company was joined back together, the wounds began to disappear. IBM became successful again. But, even so, the old IBM would never return.

eBusiness Was Gerstner's Theme Song

Indisputably, Lou Gerstner's battle cry was eBusiness, and he led IBM well in this regard. He coined the term. When he arrived in 1993, amidst all of the company's other problems, IBM was not positioned to do well in much of anything, including the Internet. Through sheer neglect, and a touch of arrogance, neither the mainframe nor the AS/400 product lines were capable of providing even a basic Internet service.

With at least a basic Internet service being a prerequisite for eBusiness, Gerstner very quickly recognized the problem and directed the solution. He announced that all of IBM's server computers would fully support Internet protocols and be prepared for eBusiness. He set up a dedicated Internet organization in IBM until things were rolling on track. It was a tall order, but IBM was able to comply. In the Fall, 1999, the Internet launch was fully completed, and Irving Wladawsky Berger, the head of the Internet unit, had his success acknowledged, and was given another challenging assignment, Linux.

Lou Gerstner was successful in having all of IBM's divisions hold at attention, ready to serve any of the needs of eBusiness. Clearly, IBM's renewed commitment to the Internet has permitted the company to survive the client/server era and begin to prosper in the new server-centric world of eBusiness and the Internet.

The Chessmaster/6000

With Gerstner at the helm, IBM became a tough competitor in all facets of the computer business, including the apparently trivial. Gerstner believed that IBM should be the best in the markets in which it chose to compete. In one of the most famous computer events of all time, in May 1997, as the world looked on in suspense, IBM dramatically demonstrated its computing potential with a machine called Deep Blue. IBM's fastest computer at the time, this 32-node IBM RS/6000 SP computer entered a chess game as an underdog for the last time. IBM had not been known for building the fastest computers. As of May, 1997 this was no longer the case.

IBM programmed this computer to play chess on a world class expert level. To test its work, IBM invited World Chess Champion Garry Kasparov to a six-game match in New York. Deep Blue, running on a machine capable of assessing the ramifications of 200 million chess moves per second, defeated Kasparov. This sent shockwaves around the world with public debate on how close computers really come to human intelligence. It also sent the Russian away hoping for a rematch... hopefully during a power storm?

Lou Gerstner's plan continued to work. Keeping the company together permitted IBM to capitalize on the new technology wave. Into the new millennium, Wall Street acknowledged that IBM was again on the right track, as the street moved the price of IBM stock back to a respectable level.

Services Are Not an Afterthought

A total solutions company must bridge the function gaps of software with high quality services to create complete solutions. Gerstner was high on bringing the total resources of IBM to its customers to solve business problems. A strong services business was necessary to round out IBM's total picture. Under Gerstner, services soon became the fastest growing segment of the company, with growth at more than 20 percent per year. All of this success had a dramatic effect on investors' perceptions, as the market value of the company increased by more than $50 billion in the short span from 1993 to 1996. Clearly, Mr. Gerstner had won public and stockholder approval.

Preparing for a New Chief

On July 26, 2000, IBM announced the appointment of Samuel Palmisano as president and chief operating officer. Lou Gerstner remained as Chairman and CEO but announced he would be retiring at the beginning of 2003. Effective March 1, 2002 Palmisano was appointed President and CEO. On October 29, 2002, the IBM board elected Palmisano to serve as the Chairman of the IBM Board effective January 1, 2003 at which time Louis V. Gerstner Jr. retired form the company.

When Lou Gerstner took over IBM, the company was faltering and in his first year, IBM revenues totaled 64.5 billion. Two years before Gerstner retired, gross revenue had climbed to $88 billion. However, then came the "recession." following 9/11. The economic downturn lasted into 2002. At the end of 2001, two months before Lou Gerstner turned the reins over to Palmisano as CEO, IBM revenues fell to $85 billion. At the end of 2004, Palmisano's IBM had stepped up revenue production to just over $96 billion. With the sale of the $12 billion PC business registering in 2005, it will be interesting to see whether IBM hits the mythical $100 billion mark that has been the goal since the John Opel days.

IBM's Chip Initiatives

Considering that IBM's RS/6000 Unix server was the first machine equipped with a PowerPC microprocessor in the very early 1990's, it is historically significant. This architecture was originally called "PowerPC" by IBM. The term "POWER" was coined several years later to differentiate between IBM's server-oriented processors and their desktop and embedded processors. For a time, IBM had two separate divisions working on POWER projects. The microelectronics area of IBM continued to call their products PowerPC based, whereas the server divisions called their processors POWERX based. The X was a number that designated the generation of the machine. For example IBM's POWER machines are at the POWER5+ level as this book is being written.

Within the last two years, IBM has merged the two divisions under one roof so that the Power Architecture now has just one IBM face.

Some synergy had been lost by the separation and it became difficult for IBM to focus on what was good for its own servers when its design decisions were being overly influenced by direct chip customers. For example the last PowerPC chip version that was built by the microelectronics division did not include circuitry necessary to run IBM's own OS/400 operating system. IBM addressed the sub-optimization problem so that now all POWER developments take into consideration all of IBM.

IBM and Apple

As noted in Chapter 5, in the early 1990's IBM approached Apple with the goal of collaborating on the development of a family of single-chip microprocessors based on the PowerPC architecture. Soon after, Apple asked Motorola to join the discussions because of their long relationship, their more extensive experience with manufacturing high-volume microprocessors than IBM and to serve as a second source for the microprocessors. This three-way collaboration became known as AIM alliance, for Apple, IBM, and Motorola. IBM continues to make chips for Apple but at some time in 2006, Apple will begin to use an unidentified Intel processor chip in its new Macs.

Game Consoles

For the past six years, IBM has had a major erelationship with Nontendo, with the first product being the Nontendo CPU, based on IBM PowerPC Technology. IBM has maintained its rrelationship with Nontendo and has been selected to provide the CPU for the coming Nintendo Revolution game console. In 2003, IBM inked Microsoft for a PowerPC derivative to drive its next generation Xbox 360. In doing this, IBM outran Intel for the business. IBM believes that it is the only company that is capable of providing what Intel needs in its next game console and the first five letters begin with POWER.

IBM's Cell Processor

Besides its Power Architecture and PowerPC initiatives, IBM also has another big game. Actually, it is a game chip that the company calls the Cell processor. With limited fabricating facilities in the industry,

IBM formed a partnership with Sony and Toshiba to assure enough supply of its now famous POWER-based game chips. IBM has taken the lead in this partnership to create derivatives of its POWER processors that can be manufactured by any of the companies and used in any of their game products. Big Blue is actually in the process of having its cake and eating it too.

The IBM Cell processor, announced to the world in February 2005 is the first product of this partnership. The Cell consists of a general-purpose POWER processor core connected to eight special-purpose digital signal processor (DSP) cores. IBM calls the DSP cores "synergistic processing elements" (SPE). These are really the heart of the entire Cell concept.

> Note DSP: Digital Signal Processor. A specialized CPU (microprocessor) used for digital signal processing.. Digital Signal Processing. DSP computers are used to process signals such as sound, video, and other analog signals which have been converted to digital form. Common uses of DSP are to decode modulated signals from modems, to process sound, video, and images in various ways, and to understand data from sonar, radar, and seismological readings.

The story on IBM's POWER technology and its probable future continues in Chapter 15.

.

Chapter 12
The Processor Chip War Is On

Intel and AMD – the Major Combatants

There is no question that the microcomputer semiconductor chip business is not an easy place to toil. Look at Intel. Overall, the company had good fortune and it continues to ride that fortune. Along the way it has established a nice chip business using its x86 architecture as the basis. But, Intel does not really have it easy. Some may say that it should but it certainly does not. On the one hand, Intel has had clone manufacturers as its major nemeses and on the on the same hand, the clone manufacturers have had the US and other governments on their side to help assure that Intel does not become the sole source for the industry.

It is not good to have the government as your adversary as IBM learned in the early 1980's when the company basically gave away its fledgling PC business to avoid the US Justice Department. (The fact that the global PC business is now well over $200 billion per year and IBM is just now approaching $100 billion, says taking it lightly in the PC marketplace was not such a good move on IBM's part) It is also not good to have a few dogs biting at your pant leg. Companies like NextGen, Cyrix, and AMD have been gnawing away at the bottom of Intel's pants for many years now. While AMD acquired NextGen and its Pentium II clone, Cyrix disappeared from real x86 competition. From three dogs to one, at least Intel's pant legs may now be in better shape, but with scrappy AMD as its major competitor, maybe not.

At the rpesent time, there are only four bona fide x86 chip makers in the marketplace, and only two of these are big time at this point. Of course, there is market leader Intel, enjoying revenues approaching $40 billion and there is the eternal challenger AMD with revenues just passing $5 billion for 2004.

What about the Non x86 Chip Makers?

There are a number of non x86 competitors in the marketplace. Two of the most famous are ARM and MIPS. Both of these companies offer single and dual processor chips and both are moving along very nicely especially in the embedded processor space.

What about Transmeta and Centaur/Via?

Besides AMD and Intel, there are two other x86 chip makers, Transmeta and Via's Centaur Technologies.

Transmeta

Transmeta opened its doors in 1995 as the chip company that was finally going to overtake Intel. In 2000, the company unveiled its crusoe line of processors that performed a "code morphing," or emulation function to be more exact, in order to host the Intel x86 instruction set, the most necessary feature of x86 compatibility. By 2005, Transmeta's "crusoe" chips were in a number of notebooks from Sony, Sharp and others, because the company had developed some nice techniques for power conservation and power leakages. However, with a market share well below 1%, Transmeta has basically cashed in its chips and the company will concentrate on selling its low-power technology to others in the industry.

Transmeta has not completely exited the x86 market however, and its chips are still made at IBM's fab in East Fishkill NY, but it will take some doing for the company to ever be a potential force with AMD and Intel, and now Centaur all being in more favorable financial positions.

Centaur Technology

Centaur Technology was founded in 1995 by Glenn Henry, a former Dell CTO and IBM Fellow. Henry had the then-revolutionary idea that a small team of microprocessor engineers could design an affordable x86 processor for the neglected sub-$1000 PC market. In ten years, Centaur has succeeded by designing the world's smallest x86 processors, shipping eleven different parts, and consistently having the fastest design cycle in the industry. Centaur is currently developing a 2 GHZ x86 processor and, in collaboration with its parent company, VIA Technologies, Centaur is positioned to tap what will soon be the biggest market in the industry - the sub-$400 PC market. Centaur's processors made headlines in 2002 with a $199 PC from Wal-Mart.com. Centaur is a wholly owned subsidiary of Via Technologies.

AMD and Transmeta are not the only x86 chip company with ties to IBM. In fact, for years IBM made Cyrix's x86 clone chips. Centaur's next-generation x86-compatible processor started production in an IBM facility in the second quarter of 2005. Via calls its C7 "the world's smallest, lowest power, and most secure native x86 processor." It targets thin and light notebooks, mini-PCs, green clients, personal electronics, and high-density servers and appliances.

In many ways, Centaur is after the same low-wattage market as Transmeta but instead of hoping to out-perform Intel, Glenn Henry and company are looking to create processor chips that are "almost as fast" as Intel but not quite. They kicker is that instead of $150.00 per Intel chip, his price is more like $30.00. That difference sure can give a PC manufacturer a lot of room to add the other components necessary for an inexpensive PC system.

According to Via, the new Centaur-created processor was designed from the ground up for low power and footprint. It is being manufactured on 90nm SOI (system-on-insulator) process technology at an IBM plant in East Fishkill, New York. Via noted that IBM's manufacturing technology "permitted unprecedented levels of integration, providing the basis for significant performance boosts within industry-leading power and thermal profiles." The new chip draws as little as 100mW (0.1W) when idle, and has a peak power rating of 20 Watts at its initial maximum clock speed of 2GHz -- 40 percent cooler than competing solutions, according to Via.

Glenn Henry Ties to IBM

As an aside, my peers in IBM often relate the role of Glenn Henry, the Centaur CEO in developing a number of IBM business processors over the years, most notably, the System/38. He is well respected in

IBM. The System/38 is the predecessor of the AS/400 and its advanced architecture still exists in the company's eServer iSeries and its new i5 models. Before his tenure at Dell, Henry served 21 years with IBM. He was the lead architect and development manager responsible for the IBM System/32, System38, and RT/PC (forerunner of Power systems).

Unlike Transmeta, Centaur looks like it will be around for awhile. Although its chip production of x86 compatibles is still well below 1% of the market, Glenn Henry's business model may pay off in the sub $400 PC market and that may be any time soon.

The Itanium -- HP & Intel

History will show that in the past ten years Intel and its partner HP spent hundreds of millions if not billions of dollars trying to move away from the Intel-created x86 architecture of processor chips. Ironically the chip style from which Intel was hoping to migrate was its very own creation and it is a style of computing that many chip manufacturers have tried to imitate.

The x86 also continues to be the major source of Intel revenue. It is the cash cow that funds all else in Intel. The giant chip maker's desire to create a super 64-bit chip that was so far ahead of all other processors began in the early to mid 1990's. The advanced nature of the chip precluded its compatibility with its own x86 chip line (Pentiums etc.)

After choosing not to use Intel's new chip in its server line, by 1995, IBM had already announced, produced, implemented, and shipped a unique 64-bit RISC-based chip built according to IBM's own Power Architecture. Intel and HP were convinced that their partnership and the design of their new super chip would match and eclipse IBM's industry leading efforts in RISC based 64-bit POWER computing.

The partners spent years developing the architecture and years in building a prototype and years in bringing the first iteration to manufacturing. All the while, they had to remember that their technology and pricing had to keep up with their competitors' at the pace of Moore's Law. Along the way, inherent complexities and bottlenecks emerged as the technology developed and matured to the implementation stage. Whether Intel and HP rose to the challenge adequately enough with their Itanium offering is for historians to debate over the years. But one thing is for sure; their Itanium initiative has yet to gain enough market acceptances to sustain it.

Success in the Chip Market

If you are a chip manufacturer-- AMD, IBM, Intel, Centaur, or whoever -- you must anticipate the ways people will use your chips and in many ways you must be willing to change to meet the constant need to change. Chip manufacturers must either predict the market and come out with a timely product, or drive the market and make consumers follow their lead.

Intel had been trying to drive the market forward for the past several years with its Itanium processor--an almost complete departure from its x86 line, and the stepchild of Hewlett-Packard's PA-RISC line. The giant chipmaker believed that its Itanium efforts would eventually replace both of these processors. However, one might suggest right processor, wrong time! If Intel were able to have brought the chip out in the 1997 or even the 1998 time frame and it performed to expectations, marketing inertia may have carried the day. If Microsoft had been able to match the 1997 delivery of the new chip with a Windows version and a working version of its office line, things would have been lots different for Itanium.

Right now, with the x86 at 64-bit and dual core already, supported on Server and Workstation OS versions, the Itanium chip is behind in obvious technology. Moreover, the chip fits squarely into the enterprise space, competing with the likes of Sun Microsystems' SPARC and IBM's well proven Power Architectures. In other words, it does not fit in the major chip marketplace that Intel has owned from time immemorial (x86) and it does not look like it is going to defeat the SPARC or Power Architecture any time soon.

While making sure it made its trip to the bank on a timely basis, to get in this situation, as great a company as it is, somebody at Intel was not paying full attention to what was going on. Fully immersed in two notions that were killing the company, Intel stuck to its guns on both until the facts were in and the company had to say "can't do." First Intel was pushing the Itanium to be accepted by a market in which it did not fit. Second, Intel did not support the notion of a next generation 64-bit processor chip with x86 32-bit backwards compatibility. Only a company with such huge capital and market clout would be able to make these two major marketing blunders and still continue as the force with which to reckon in its industry.

Intel Itanium Helps AMD

While Intel played the stubborn leader game, somehow challenger AMD was able to develop and offer the market exactly what it wanted and needed – 32-bit and 64-bit x86 computing on the same processor chip. AMD took a lesson from partner IBM's POWER chips in which for many years; Big Blue had successfully placed the full instruction sets of multiple computer architectures on one chip.

AMD First to 64-bits

AMD dubbed its new x86 technology extensions as "AMD64." Intel of course owns the x86 instruction set and AMD gets to use it by a zero fee licensing arrangement with Intel that the company won in its longstanding lawsuit several years back. Yet it was AMD as the "lessee," not Intel as the "lessor" that in the second quarter of 2003 developed a way to keep the x86 instruction set alive and well with its new Opteron server and later with its Athlon-64 workstation line.

AMD beat Intel to market and made the first x86 chips to include enterprise-oriented accommodations for 64-bit computing while offering full backward compatibility for 32-bit programs. AMD not only beat Intel to the general marketplace of 64-bit computing, the company added more goodies to the chips enabling vastly improved performance by, among other things, increasing the number and size of registers in the processor.

First and Fastest

Unlike the Itanium's notoriously slow 32-bit performance, AMD's Opteron and Athlon-64 work naturally in 32-bit mode, making migration ports to 64-bit applications necessary only in very rare situations, such as when memory requirements exceed the 32-bit limit. In almost all other situations, the old applications just run in 32-bit mode on the new AMD chips. To do both 64-bit computing and 32-bit computing with Intel at the time, you needed two machines, a 64-bit Itanium and a 32-bit Pentium.

AMD shocked the industry by defining a number of new instructions to the x86 instruction set to support 64-bits. Moreover, AMD talked Microsoft into using their instructions in an upgraded 64-bit Windows so it was already too late for Intel to counterpunch. When AMD was able to do all of the old Intel 32-bit stuff and new 64-bit stuff with its new chip, and even Windows worked, Intel was nothing more than flabbergasted.

Intel is still not ready to abandon its Itanium, yet it has been consistently losing in the court of public opinion. To get back on track and stop the sales and the PR hemorrhage from AMD's accomplishments, in 2004, a year after AMD64 took over its world, Intel released a version of its always powerful Xeon processor using a version of AMD64 that it had reverse engineered. In other words, Intel used the AMD designed 64-bit instructions to give its customers what they were asking. Intel called the "new" architecture EM64T. The computer geeks and nerds on the Internet call it AMD64 reverse engineered by Intel.

Intel Reluctant to Be Successful

The Xeon processor that the company selected had been code named internally as *Nacoma* and because Intel still did not fully embrace the AMD64 style of computing, they released the new 32-64 bit chip with little fanfare. Basically, Intel quietly and one would have to conclude reluctantly, adopted (cloned) the AMD64 architecture so that its chips could participate in the AMD-fostered 64-bit revolution. There is no question that a company as rich in R&D as Intel could have engineered and built its own AMD64 architecture at any time during the ten years that passed since IBM introduced its 64-bit computing model with POWER in 1995. IBM chose to make its own chips rather than buy Itanium and so far that decision looks like a good one for IBM.

Indeed, AMD forced Intel to pit its own CISC-based Xeon processor, in 64-bit mode, against its 64-bit VLIW-based Itanium. Instead of upgrading a 32-bit Xeon-based server to the Itanium or to a RISC-based system, for example, Intel x86 Xeon class users were able to simply switch to the next level of 32-64 bit Xeon x86.

X86 on Intel's Front Burner

Even now, it appears that Intel is not fully revealing its plans for the future of its Itanium architecture but one thing is for certain. The 15 new processors in the Intel pipeline are not Itaniums. That sure says something. It says that because it will be very hard to get the Itanium back into the front line, Intel may be forced to keep the processor as a side line, waiting for its day to come.

With the AMD wake-up call, Intel is now very tuned into market momentum. Its own x86 architecture is being used for more and more enterprise-oriented tasks--Web, database and computing chores that were once reserved for beefier architectures. The EM64T based single and dual core processors can now handle big workloads in either the Linux environment or 64-bit Windows. That is not bad at all for Intel.

The giant chipmaker chose to take a big chance with the Itanium and it had been trying to steer the market along its path. So far Intel has had great difficulty in these efforts. Meanwhile, the dog at its pant leg (AMD) single handedly has breathed new life into Intel's own x86 architecture. AMD proved that it was willing to take risks in exchange for opportunity, and it proved that the company is a worthy competitor far beyond the consumer market segment.

The fact is that many enterprises have found it more convenient to use the same hardware architecture for both low-end and high-end applications. If a company can save money by going all x86, history says that's the route they'll take. Even with its enormous resources, trying to swim upstream with a big fat "Itanic" in its jowls, even at a competitive price, has gotten Intel nowhere. EM64T is Intel's tacit acceptance of that reality.

The Giant Shows Agility

As noted in Chapters 9 and 10, in April 2005 both AMD and Intel entered the multi-core processor chip arena with their dual core offerings of take-home processor chips. Through its press releases, it was very easy to discover that AMD was very upset that Intel had tuned into the AMD announcement cycle and out of nowhere announced its own dual core models three days before AMD's already scheduled product launch.

It is unusual for the giant to show agility, but can you blame Intel for acting like a clone manufacturer? After choosing for ten years not to build a chip that would give its customers access to their existing 32-bit Windows and/or Linux software along with powerful 64-bit facility and dual core technology, Intel took notice that AMD had solved that problem for AMD customers as well as Intel's customers. And, then, with a little chicanery, they beat AMD to market by three days with their dual core ground breaking solution.

Go Where the Customers Want You to Go

Losing customers has never been Intel's game. Once the notion of the AMD64 hit Intel, they knew that it was not the technology as much as the design and the idea that made it a market winner. With all the Itanium dollars and Intel's massive R & D budget, it seemed unlikely that Intel would have acceded to the wishes of its customers if AMD had not beaten them to the punch. Again, Intel could have had its own solution ten years ago but apparently chose not to.

The Itanium project prevented the company from satisfying its own customers' wants and needs. So, without the need to acknowledge the old cloner, AMD, Intel merely figured out how AMD did its AMD64 thing and copied it. Though that is not what we would expect from the undisputed leader in x86 sales, it did enable Intel to quickly announce its own version of AMD64, announced as EM64T -- the Intel 64-bit architecture for non-Itanium class chips.

Has Intel Gotten the Message?

So, where does that leave us? Some might say if Intel actually brought its Itanium development team back to the x86 platform, there would be an opportunity for phenomenal innovation. But, even in 2005, Intel is still not ready to publicly give up the ghost though privately, the ghost of Itanium may already be gone.

None of the new processor chips that I have reviewed on the Intel announcement block are substantially different than what can be bought today. So, will Intel wait for the next major AMD innovation to react or will Intel get in front of the crowd with a series of processor chips that provide substantially better x86 technology and 64-bits with no buts? The greatest fear that AMD must have is that Intel, after being burned by AMD, may now actually understand its market as well as AMD understands it. It would not be a good AMD omen for Intel to stay agile and alert for the company may begin to innovate for the market, and with all of the chip giant's technology soldiers, along with its massive R&D, the Intel Corporation can very well be a formidable foe.

Is there a new market and opportunity-driven Intel for AMD to deal with? It will be hard for AMD to beat Intel if Intel chooses to use Jerry Sanders' proven techniques to fight AMD's innovations. On the other hand, the more Itanium news you see, the more it means that Intel's show of agility and customer awareness was not a change of heart but a knee-jerk reaction to having lost a big technology battle. If this is the case, there will be many future opportunities for AMD to gain market share.

Chapter 13
The War Escalates– AMD v. Intel

AMD v. Intel

Over the years AMD has had its court battles with Intel and for the most part, the company has won the day. One of the reasons that AMD can compete so well against Intel is that in 1991, the chip maker won an unlimited license to the x86 instruction set in a major lawsuit. After 5 years of litigation an arbiter awarded AMD $10 million as well as a permanent, nonexclusive and royalty-free license to any Intel intellectual property embodied in AMD's own 386 microprocessor, including the x86 instruction set.

Buoyed with this win, AMD set about re-inventing itself. And, so, AMD is still alive and still feisty enough to think it can win against the Intel juggernaut. Time will tell but it seems that AMD thinks it needs some more help from Uncle Sam and from other regulatory bodies across the world in order to be able to be treated fairly by Intel.

On the one side of this battle, we have the champion who got to where it is, the largest, bar none, manufacturer of processor chips in the world, by using some great technology and some great luck with its IBM contract in 1981. On the other side, we have the contender, the scrappy Sanders-built company, 1/8 the size of the leader but with a heart equal in size or greater.

In March, 2005, the scrapper from Sunnyvale got real upset with what was happening to its worldwide x86 marketplace. After clearly having the best technology for the past several years, AMD alleged that it could not get even its loyal customers to buy from the company even though they wanted to. AMD believed that Intel was behind their marketing problems. Why else would the best technology not sell?

So, when the Fair Trade Commission of Japan (JFTC) ruled that Intel had violated the anti-monopoly act, AMD was not surprised. They believed that they had been living it. Shortly after the ruling, Intel responded publicly and from AMD's perspective in its response, the Chip Giant had refused to acknowledge the JFTC facts and the harm that had been caused to competitors and consumers alike.

On March 31, in frustration, AMD responded to Intel's "Acceptance" of The Fair Trade Commission of Japan's ruling of Intel's violation of Section 3 of Japan's Antimonopoly Act by releasing the statement which follows:

"It is unfortunate that even when presented with specific – and very disturbing – findings of deliberate and systematic anti-competitive behavior, Intel refuses to face the facts and admit the harm it has caused to competitors and consumers," said Thomas M. McCoy, AMD executive vice president, legal affairs and chief administrative officer. "Although Intel's willingness to comply with the JFTC Recommendation is a step in the right direction, it has conspicuously failed to either accept responsibility for its actions or acknowledge that competition is best served when customers and consumers have a choice. The JFTC determined that Intel conditioned its pricing based on customers not doing business with competitors; governments around the world must ensure that such anti-competitive actions are not impacting their markets as well."

Issued on March 8, 2005, the JFTC Recommendation found that Intel had abused its monopoly power to exclude fair and open competition, thereby violating Section 3 of Japan's Antimonopoly Act. These findings reveal that Intel used illegal tactics to stop AMD's increasing market share by imposing limitations on Japanese PC manufacturers. Specifically, the JFTC found that:

Some manufacturers were required to buy 100 percent of their CPUs from Intel; another manufacturer was forced to curtail its non-Intel purchases to 10 percent or less;

Intel separately conditioned rebates on the exclusive use of Intel CPUs in specific PC product lines or brands in order to eliminate competitor CPUs from key OEM brand lines;

Intel's activities in Japan include the use of its "Intel Inside" program, market development funds and other rebate programs. Intel's tactics lock-in OEMs and prevent them from doing business with Intel's competitors.

The Recommendation reveals that Intel imposed these restrictions in direct response to AMD's growing market share in Japan from 2000-2002.

The Recommendation also notes that as a result of this misconduct, the combined Japanese market share of AMD and a second, much smaller CPU company fell from 24 percent in 2002 to 11 percent in 2003.

The JFTC imposed a number of restrictions on Intel including ending the use of rebates and other funds to illegally distort competition, notifying its customers and educating its employees that it may no longer provide rebates and other funds to Japanese computer manufacturers on conditions that exclude competitors' CPUs.

Intel had ten days to either comply with or appeal the ruling and the company requested a two week deadline extension. After the two weeks, Intel elected to accept the recommendation, which means that the company will need to alter business practices and implement other remedies immediately to meet the guidance in the recommendation.

There is more bad news for Intel and perhaps good news for AMD. The European Commission recently stated that it is investigating Intel for possible similar anti-competitive business practices in Europe and is cooperating with the Japanese authorities.

Writing It Down Makes It Better Sometimes

At the time I read AMD's statement, it reminded me of what my friends tell me when I am upset about something and I am ready to write a nasty letter to someone about how I feel. They tell me to write the letter but don't send it right away. Sit on it. Consider it, and then decide what to do. I wondered what AMD was going to do, obviously feeling wronged by Intel and feeling a sense of vindication by the

JFTC ruling. They had issued the letter in a press release, but was that all there would be? Three months later, almost to the day, I got my answer.

AMD Files Antitrust Complaint against Intel in U.S. Federal District Court

On June 27, 2005, AMD took its battle to the courts as it filed an antitrust complaint against Intel Corporation ("Intel") in U.S. federal district court for the district of Delaware under Section 2 of the Sherman Antitrust Act, Sections 4 and 16 of the Clayton Act, and the California Business and Professions Code. The 48-page complaint is written in terms that do not demand computer literacy. It is actually good reading for anybody who wants to understand one of the latest major volleys in the Chip Wars. You can get to it at http://www.amd.com/breakfree.

The complaint explains in detail how AMD views Intel's practices, many of which AMD complained about in its March 30 press release as discussed above. The complaint details AMD's allegations of worldwide coercion of customers and it also asserts that Intel's "illegal acts" have inflated computer prices and have limited choices for businesses and consumers.

AMD alleges in its complaint that Intel has unlawfully maintained its monopoly in the x86 microprocessor market by engaging in a worldwide coercion of customers from dealing with AMD. It identifies 38 specific companies in various positions of the supply chain that have been victims of coercion by Intel. The list includes large scale computer-makers, small system-builders, wholesale distributors, and retailers, through seven types of illegality across three continents.

"Everywhere in the world, customers deserve freedom of choice and the benefits of innovation – and these are being stolen away in the microprocessor market," said Hector Ruiz, AMD chairman of the board, president and chief executive officer. "Whether through higher prices from monopoly profits, fewer choices in the marketplace or barriers to innovation – people from Osaka to Frankfurt to Chicago pay the price in cash every day for Intel's monopoly abuses," according to Ruiz..

From reading this book we all know that x86 microprocessors run the Microsoft Windows, Solaris, and Linux families of operating systems. Even Apple, a pioneer of the PC and one of the industry's enduring innovators, announced that it would switch exclusively to x86 processors to run its Mac OS software beginning in 2006. Intel's share of this critical market currently counts for about 80 percent of worldwide sales by unit volume and 90 percent by revenue, giving it entrenched monopoly ownership and super-dominant market power. x86 is everywhere and Intel "owns" the x86 way.

AMD is Feeling Pain

You can tell by the tone of its press releases that AMD is feeling pain and it has taken the gloves off to help gain public support. AMD's comments to the press are anything but kind regarding its actions against Intel.

"You don't have to take our word for it when it comes to Intel's abuses; the Japanese Government condemned Intel for its exclusionary and illegal misconduct," said Thomas M. McCoy, AMD executive vice president, legal affairs and chief administrative officer. "We encourage regulators around the world to take a close look at the market failure and consumer harm Intel's business practices are causing in their nations. Intel maintains illegal monopoly profits at the expense of consumers and computer manufacturers, whose margins are razor thin. Now is the time for consumers and the industry worldwide to break free from the abusive Intel monopoly."

Points in the Complaint

According to the complaint, Intel has unlawfully maintained its monopoly by, among other things:

- Forcing major customers such as Dell, Sony, Toshiba, Gateway, and Hitachi into Intel-exclusive deals in return for outright cash payments, discriminatory pricing or marketing subsidies conditioned on the exclusion of AMD;

- ✓ According to industry reports, and as confirmed by the JFTC in Japan, Intel has paid Dell and Toshiba huge sums not to do business with AMD.

- ✓ Intel paid Sony millions for exclusivity. AMD's share of Sony's business went from 23 percent in '02 to 8% in '03, to 0%, where it remains today.

- Forcing other major customers such as NEC, Acer, and Fujitsu into partial exclusivity agreements by conditioning rebates, allowances and market development funds (MDF) on customers' agreement to severely limit or forego entirely purchases from AMD;

- ✓ Intel paid NEC several million dollars for caps on NEC's purchases from AMD. Those caps assured Intel at least 90% of NEC's business in Japan and imposed a worldwide cap on the amount of AMD business NEC could do.

- Establishing a system of discriminatory and retroactive incentives triggered by purchases at such high levels as to have the intended effect of denying customers the freedom to purchase any significant volume of processors from AMD;

- ✓ When AMD succeeded in getting on the HP retail roadmap for mobile computers, and its products sold well, Intel responded by withholding HP's fourth quarter 2004 rebate check and refusing to waive HP's failure to achieve its targeted rebate goal; it allowed HP to make up the shortfall in succeeding quarters by promising Intel at least 90% of HP's mainstream retail business.

- Threatening retaliation against customers for introducing AMD computer platforms, particularly in strategic market segments such as commercial desktop;

- ✓ Then-Compaq CEO Michael Capellas said in 2000 that because of the volume of business given to AMD, Intel withheld delivery of critical server chips. Saying "he had a gun to his head," he told AMD he had to stop buying.

- ✓ According to Gateway executives, their company has paid a high price for even its limited AMD dealings. They claim that Intel has "beaten them into 'guacamole'" in retaliation.

- Establishing and enforcing quotas among key retailers such as Best Buy and Circuit City, effectively requiring them to stock overwhelmingly or exclusively, Intel computers, artificially limiting consumer choice;

- ✓ AMD has been entirely shut out from Media Market, Europe's largest computer retailer, which accounts for 35 percent of Germany's retail sales.

- ✓ Office Depot declined to stock AMD-powered notebooks regardless of the amount of financial support AMD offered, citing the risk of retaliation.

- Forcing PC makers and tech partners to boycott AMD product launches or promotions;

- ✓ Then-Intel CEO Craig Barrett threatened Acer's Chairman with "severe consequences" for supporting the AMD Athlon 64™ launch. This coincided with an unexplained delay by Intel in providing $15-20M in market development funds owed to Acer. Acer withdrew from the launch in September 2003.

- Abusing its market power by forcing on the industry technical standards and products that have as their main purpose the handicapping of AMD in the marketplace.

- ✓ Intel denied AMD access to the highest level of membership for the Advanced DRAM technology consortium to limit AMD's participation in critical industry standard decisions that would affect its business.

- ✓ Intel designed its compilers, which translate software programs into machine-readable language, to degrade a program's performance if operated on a computer powered by an AMD microprocessor.

No Marketing Clout

As we have already noted in this book, AMD is not an also-ran in x86 technology. It is now the technology leader. Unfortunately for AMD, the company does not exert major market clout to help its potential customers know that they would be OK supply wise with AMD. AMD's fabs, for example, may not be able to make enough chips to satisfy market demand if the whole world were to go AMD tomorrow, but they could outsource to other fabs. There is enough fab space out there in companies like Chartered Semiconductor and others that AMD chips could be coming off a production line to meet demand faster than you can say Intel. AMD is a member of the same consortiums as IBM and there are a number of fabs who would be ready to help AMD if supply became the problem. However, AMD PC makers cannot say "Intel Inside" on its units.

If AMD's allegations are true, then with sufficient manufacturing capacity, theoretically AMD could supply the whole industry and Intel, without Uncle Sam's help would have to compete fairly. Well, not exactly. Intel makes more and more of the parts that go into a PC. The giant chipmaker is also a giant circuit board maker and a giant compiler provider and it also has other parts upon which its customers depend.

An example is the Centrino assembly for laptops in which Intel makes most of the product's insides. Moreover, the company has plans for things like Desktrino in which Intel supplies a bundle including the chip for desktop makers. If PC, laptop, and server makers start buying these bundles with Centrino Inside, Desktrino Inside, etc., one thing is for sure. There will be no AMD chips in the bundles, unless the courts decide that there will be.

Technology Isn't Everything

Leading publications such as The Wall Street Journal, The Washington Post, The Economist, San Jose Mercury News and CNET have in recent years all recognized AMD as a leader in microprocessor innovation. AMD has achieved technological leadership in critical aspects of the x86 market, particularly with its AMD Opteron microprocessor, the first microprocessor to take x86 computing from 32 to 64 bits, and with its dual-core processors. In today's world, however, that may not matter if Intel can keep AMD's prospects from buying from them.

For its money, Intel is not buying AMD's arguments. In prepared statements on June 28, Intel countered AMD's charges: "We strongly disagree with AMD's complaints about the business practices of Intel and Intel's customers. Intel believes in competing fairly and believes consumers are benefiting from this vigorous competition. Intel will respond appropriately to AMD's latest complaints and is committed to successfully resolving these issues in court."

AMD Sues Intel in Japan

In another salvo against Intel in the world community, on June 30, 2005, AMD Japan Filed a claim for damages arising from the violations of Japan's Antimonopoly Act against Intel K.K. in Tokyo High Court and Tokyo District Court. In March the JTFC ruled against Intel and Intel accepted its ruling. AMD is about to see just how far that acceptance takes it. From AMD's perspective, AMD Japan sought to restore fair and open competition for the Japan PC and server marketplace and that has not happened.

Intel did not immediately respond to the June 30 filing in Japan but CEO Paul Otellini did offer a statement on June 29 regarding the US suit that clearly shows Intel's posture towards AMD in both actions:

"Intel has always respected the laws of the countries in which we operate," Otellini said. "We compete aggressively and fairly to deliver the best value to consumers. This will not change.

"Over the years, Intel has been involved in other antitrust suits and faced similar issues. Every one of those matters has been resolved to our satisfaction. We unequivocally disagree with AMD's claims and firmly believe this latest suit will be resolved favorably, like the others."

The Chip War Is On

Not only do AMD and Intel disagree on Intel's business practices but the cases that Otellini cites above appear in the victory column for AMD on its Web site. Even long proven facts are interpreted differently by the two combatants. They don't agree to disagree, they just diaagree. One thing is for sure, there is a real chip war going on between AMD and Intel and AMD may be about to get some help from its friends.

Chapter 14
IBM: The Quiet Semiconductor Technology Leader

Over Forty Years and Stronger than Ever

IBM has been innovating in semiconductor designs since long before the System/360 mainframe computer was introduced in 1964. Though IBM has never really been known as a component manufacturing company, the company almost bet the farm on technology with its introduction of the System/360 as it created one plant after another to be able to provide the chips for its own systems. Since then and before, IBM has been a leader in the production of semiconductor technology holding more patents than all of its competitors combined.

As a sampling of the armaments IBM has ready to go in its chip war chest, this chapter highlights five specific areas in which IBM patents have born fruit:

1. Silicon Germanium
2. Copper in semiconductor chips
3. Silicon on insulator
4. Low-k dielectric
5. Strained silicon with Dual Stress Liner

Silicon Germanium

Silicon germanium technology (SiGe), perfected for commercialization by IBM's Dr. Bernard Meyerson and company has become the driving force behind the explosion in low-cost, lightweight, personal communications devices like digital wireless handsets. These devices, which operate at extremely high frequencies, are seeing continuous improvements as a result of SiGe technology. Due to its highly efficient power usage, SiGe extends the life of wireless phone batteries. You can think of it as more zip per chip.

Silcon Germanium is in demand today, and IBM is a leader in the technology. In addition to its chip fabs, IBM offers service packages that provide the engineering support during the design; manufacturing; bond, assembly, and test phases of semiconductor development and production of SiGe devices. In other words, companies with an idea for the use of SiGe, can come to IBM to make the idea a product.

This IBM invention plays a significant role in the development of entertainment and information technologies like digital set-top boxes, direct broadcast satellite (DBS), automobile radar systems, and personal digital assistants (PDA). SiGe's advantages allow these popular communications technologies to be smaller and more durable.

The public will see even more benefits from SiGe technology as it works behind the scenes, as the heart of the fiber-optic lines, switches, and routers that are crucial to a networked world.

A leader in SiGe process and product technology, IBM has been at the forefront of SiGe technology since 1982, and was the first company to broadly manufacture SiGe technology. Having perfected this exceptionally high-speed, innovative chip technology, IBM has moved it from the laboratory to the marketplace. Many products built with IBM's patented SiGe process are now available.

Big Blue has its investments in place and it believes the small, multifunction, low-cost, mobile client devices, capable of communicating over voice and data networks, are a key element of the future of computing.

Copper

For ages, man has known that Copper is just about the best conductor of electricity on earth. Yet, in 25 years of trying, no one had produced a marketable copper chip...until 1997 at IBM.

Fulfilling a dream of several decades, IBM introduced a technology that allows chip makers to use copper wires, rather than the traditional aluminum interconnects, to link transistors in chips. This advance gave IBM a significant lead in the race to create the next generations of semiconductors.

Why Copper?

Every chip has a base layer of transistors, with layers of wiring stacked above to connect the transistors to each other and, ultimately, to the rest of the computer. The transistors at the first level of a chip are a complex construction of silicon, metal, and impurities precisely located to create the millions of minuscule on-or-off switches that make up the brains of a microprocessor. Breakthroughs in chip technology have most often been advances in transistor-making. As scientists kept making smaller, faster transistors, and packing them closer together, the point of interconnect started to present problems.

In the semiconductor industry, bigger isn't always better. In fact, the phrase "smaller, faster, and cheaper" is more applicable to computer technology. Aluminum had long been the conductor of choice, but it was destined to reach its technological and physical limits. As pushing electrons through smaller and smaller conduits becomes harder to do, aluminum just isn't fast enough at these new, smaller sizes.

Scientists had seen this problem coming for years and tried to find a way to replace aluminum with one of the three metals that conduct electricity better: copper, silver, or gold. Of course, if it were simple, it would have been done a long time ago. None of those metals is as easy to work with as aluminum in ever-decreasing amounts. Any new material presents fresh challenges, and reliably filling submicron holes and channels is a bit like filling the holes of a golf course from an airplane. What's worse, those metals interact badly with silicon, soaking into it and altering its electrical properties. Not by much, but even a few stray atoms are enough to short-circuit the chip. IBM had to develop a diffusion barrier that could be deposited in silicon wafers along with the copper. By the late 1980s, IBM researchers found one metal that did the trick, paving the way to the breakthrough announced in 1997.

Silicon on Insulator

Along with IBM's leadership in manufacturing microchips using copper interconnect technology, Big Blue has now developed the first commercially-viable implementation of silicon-on-insulator (SOI). IBM is the first company to be able to apply SOI in building fully-functional mainstream microprocessors, the most complex type of chip.

For more than 30 years, scientists have been searching for a way to enhance existing silicon technology to speed computer performance. This new IBM success in harnessing SOI technology results in faster computer chips that also require less power--a key requirement for extending the battery life of small, hand-held devices that have become pervasive and will be even more pervasive in the future. SOI is a major breakthrough because it advances chip manufacturing one to two years ahead of conventional bulk silicon.

SOI technology improves performance over existing technology by 25 to 35%, equivalent to two years of technology advances in the existing CMOS area. SOI technology also brings power use advantages of 1.7 to 3 times. SOI should eventually replace bulk CMOS as the most commonly used substrate for advanced CMOS in mainstream microprocessors and other emerging wireless electronic devices requiring low power.

IBM started SOI research in earnest in the mid-1980s, when researchers concluded that it would be difficult to scale bulk silicon-based circuits down to low voltages. The effort picked up momentum in 1990, when IBM engineers demonstrated that a partially depleted CMOS (rather than fully depleted) could overcome the short-channel effect common to fully depleted devices in SOI wafers.

A number of industry analysts noted that IBM's lead in copper interconnects and in SOI is "a really big deal" that puts pressure on Intel. Using copper is a really big deal, and IBM got there well before Intel. IBM engineers repeated that with SOI, and the combination of both copper and SOI results in measured performance gains equal to almost a whole process generation. When IBM rolled out these technologies at a 0.18-micron level, the company was able to apply copper interconnects and SOI-type transistors and most analysts agreed that IBM had taken major risks and for its efforts had achieved " pretty big advantage." Today, IBM is the unquestioned leader in SOI and copper interconnect technology..

Low-k Dielectric

IBM also developed a manufacturing technique known as "low-k dielectric." This technique meticulously shields millions of individual copper circuits on a chip, reducing electrical capacitive loading and "crosstalk" between wires that can hinder chip performance and waste power. The company is designing custom chips that meet the high-performance and low-power consumption demands of next-generation microprocessors and other logic chips.

Today, designers work to improve chips by adding more circuits and packing them closer together on a single piece of silicon. They must constantly battle against interconnect "parasitics" (resistance, capacitance, and crosstalk) that impact signal speeds, power consumption, and operational limits. Just as crosstalk can occur on telephone lines, chip crosstalk means that a chip is basically unusable.

IBM has made breakthrough changes in interconnect materials to stave off these limits, first by migrating from aluminum to copper conductors in 1998, and recently by changing the insulator around the conductors from fluorosilicate glass to organosilicate glass ("SiCOH"), a patented process that IBM and others have pioneered. Each change in materials has reduced interconnect parasitics by 20-40%. IBM leads the industry in the performance and reliability of these implementations.

To speed the introduction of products based on this manufacturing process, IBM has a custom chip offering called Cu-08. This application specific integrated circuit (ASIC) template will be manufactured with IBM's 90-nanometer process technology, resulting in chip features as small as 0.08 microns (more than 1300 times thinner than a human hair). Cu-08 supports designs up to an unprecedented 72 million "gates," or circuits. IBM is the world's largest ASIC producer and widely recognized as the leader in innovative semiconductor manufacturing technologies.

Improved Strained Silicon Using Dual Stress Liner

In 2003, IBM signed a major technology sharing agreement with AMD. In 2005, the companies re-upped the agreement for another three years. This agreement has already resulted in some joint achievements in chip building. In late 2004, for example, AMD and IBM announced the joint development of an advanced strained silicon technology called "Dual Stress Liner" (DSL) for use in their upcoming processor lines. The DSL technology is compatible with SOI (Silicon On Insulator) technology, but provides superior performance to the current generation of SSDOI (Strained Silicon Directly On Insulator), the combination of Strained Silicon and Silicon On Insulator.

Incorporation of the new strained silicon on production wafers was implemented in the first half of 2005 in AMD's 90 nm AMD64 devices and IBM's 90 nm POWER chips. AMD is claiming a 24% increase in speed, at similar power levels, of transistors using this technology compared to those that do not.

Interestingly, Intel has opted to throw some mud on this great news from the SOI chip makers. Intel has stayed with bulk CMOS as its preferred method and the company announced that its engineers recently developed a new strained silicon process for use in its upcoming 65 nm process. Intel claims that its process reduces leakage by 30%, but that is not directly comparable to AMD's claim of 24% increase in "speed" at similar dissipation rates. One could imagine, though, that a 30% decrease in leakage would probably allow a 24% increase in switching speed. So has Intel cloned another process to make these new strained silicon methods identical? The world may never know.

IBM thinks it knows and apparently Intel thinks it knows. IBM thinks its engineers are right on, having been dealing with heat issues from way back to water-cooled mainframe days and quality issues from the time the company switched its mainframes from regular transistors to semiconductor chips. IBM thinks its people are smarter than Intel's and so far most others in the industry, including AMD have lined up on IBM's side.

Intel has not been doing well in most of its recent bets. A look at Intel's very recent major issues in its front-line business says there is a major level of confusion within the company. Take the clock speed hoax in the initial Pentiums, or the three-year heat problem after promising a 4GHz Pentium 4 line, or Intel's reluctance to work on chip parallelism instead of just ramping up the clock speed, or the company's inability to read the market for a 32-bit compatible 64-bit chip (AMD64). All of these are Intel suppositions gone wrong. Add to this Intel's recent insistence that the older bulk CMOS technology is better than SOI plus the company's insistence that its coming depleted SOI technology is better than the IBM developed methodologies covered in this chapter. Will these suppositions hold true?

For example, with regard to just two modern day semiconductor issues, something called *the history effect* with SOI technology and the notion of fully depleted SOI as used in analog circuits, IBM fellow Ghavam Shahidi stated that the *history effect* has been overblown. "The history effect is less for fully depleted SOI, but it is a very small effect to begin with, and it's hard to measure. I'm not sure why Intel is making it a big deal." Regarding analog with fully depleted SOI, "… analog circuits still need to make contact with the [chip] body, which can't be done with fully depleted designs. I think this shows [Intel's] lack of experience," Shahidi said. You make the call.

Summary

IBM's advances in semiconductors go back a long way even before Intel was a company. It's most recent surge of innovations began in 1997. IBM engineers improved the connections between transistors by allowing copper (a better conductor of electricity) to be substituted for aluminum. Another IBM breakthrough came in 1998, when IBM turbocharged transistor technology with its unique silicon-on-insulator process that allows chips to run faster. That same year, IBM became the first company to mass produce chips made of silicon germanium to speed communications products.

Then, in 2000, IBM unveiled a new manufacturing technique that uses a material known as a "low-k dielectric" to effectively shield millions of individual copper circuits on a chip, reducing electrical " crosstalk" between wires that can hinder chip performance and waste power. Additionally, IBM brought strained silicon and more recently strained silicon with DSL to market for use in fabs that use the silicon on insulator process.

Though IBM does not appear willing to be a big time player in the foundry (fab) business per se, the company is clearly the leader in chip technology. IBM makes its best money in this industry from its R&D, however, not from its manufacturing. Since the Gerstner years, IBM has licensed the rights to use its advanced processes to companies like AMD and Intel as well as other companies that are in the Chip fab business.

IBM is not interested in a war with Intel over chip technology though clearly the two compete for world opinion. However, with its advanced semiconductor process technologies and its own POWER chip line, and its forthcoming highly advanced Cell processor, there is no doubt that IBM and Intel are going to tangle in the near future in one way or another. When that happens, AMD is clearly positioned, with its agreements and its wholehearted adoption of IBM technologies that it will line up in IBM's camp. The question is: "with its superior technology and processes, will IBM be able to win the market?"

With IBM clearly lined up as the sole chip supplier in the home and game console market, a market expected to subsume the PC market in the future; my answer to that question is a qualified "Yes!" The only qualification I have is – if IBM really wants it!

Chapter 15 IBM Power Architecture and Processors

IBM Is Far from Being a Chip Newbie

Nothing really matters in business except that which can be sold. IBM has always tuned into this simple fact of business. That is why the company's recent technology and marketing moves regarding its chip design and fabrication capabilities are most worthy of note.

In doing research for this book, I contacted Intel and AMD about their ongoing chip wars. I also contacted IBM, a company that is flying so far under the chip wars radar that the company actually can get away with suggesting that it is not engaged. "That's a war between AMD and Intel," IBM said.

Yet, IBM has a highly successful chip family today that the company refers to as its Power Architecture. IBM also makes a number of x86 compatible chips for a number of vendors at its fab facility in East Fishkill, NY. The Power Architecture has been around since the early 1990s so it is well proven, stable and mature enough to be a big threat to Intel if IBM chose to use it as a weapon. Yet, again IBM says that it is not interested in engaging Intel's x86 technology with a battle to place PowerPC or POWER on the desktop. Even still, that does not take IBM completely out of the chip wars and you are about to learn why.

Power Everywhere

On December 2, 2004 IBM unveiled its new notion of Power Everywhere and quickly trademarked it. . That's power as in Power

Architecture. At the event, IBM also announced its new Power.org consortium, consisting initially of 15 companies that will participate in the "open standards community around chips and systems which use Power Architecture technology." A partial list of companies includes Novell, Red Hat, and Sony.

eClipz

At the core of IBM's PowerEverywhere strategy, is a notion that IBM has dubbed eClipz. It is a project that is to deliver a single Power Architecture to everything from embedded systems to mainframe computers. eClipz derives its name from IBM's eServer, iSeries, pSeries, and zSeries, which for the first time will all run on the same processor. Of course if Microsoft does a good job with its Xbox 360, IBM's eServer may also have a version running on POWER, driven by Windows. Maybe that would be the wSeries.

Central to eClipz is IBM's forthcoming POWER6 processor and its many derivatives. While not specifically part of eClipz, it is expected that IBM will deliver an ultra-lite derivative of the processor, P6UL. Originally anticipated to find its way into Macs, this little guy may be the chip that provides IBM a way to compete with Intel on the desktop. The P6UL is not expected to be out until some time in 2007. IBM has said that it is currently targeting 6GHz for the first iterations of the POWER6, which will be manufactured using a 65nm process.

Power.org

At the heart of the PowerEverywhere movement is Power.org. IBM reports that the mission of Power.org is to develop, enable and promote Power Architecture technology as the preferred open standard hardware development platform for the electronics industry and to administer qualification programs that optimize interoperability and accelerate innovation for a positive user experience.

Some of the outputs expected from Power.org include the following:

Open standards and specifications
Business guidelines documents
Best practices and education
Certifications to validate implementations and drive adoption

Interoperability between community members
Sustainability built on driving open standards and convergence

To further place the Power Architecture in the heart of the technical community, on March 31, 2004, IBM formally announced the "opening" of the Power Architecture, which is a combination of expansive licensing and a set of freely available tools for designing POWER-based microprocessors.

A Community of POWER Hardware Users

Just like Linux and Java, at the hardware level Power Architecture is expected to be embraced by the hardware technical community and is now positioned to be the chip of choice for many companies who rely on generic computing power at the chip level. IBM clearly is interested in building a community that will openly collaborate and innovate around its POWER microprocessor architecture. The chip is already used in a vast range of products from the world's most powerful enterprise systems and supercomputers to games and embedded devices. Industry analysts see the PowerEverywhere move as having major implications for computers and the electronics industry at large for the long haul.

Use Our Chip; Make Your Own Product

IBM sees the move as being designed to create a platform for innovation that enables researchers and electronics makers to add the features and capabilities that will drive new devices and applications using Power Architecture. The move recognizes the fact that it is the final chip designs -- more than the underlying architecture -- where innovation is taking place to create entire systems in silicon, not just electronics components. IBM's clients will be the designers of the final products with Power as the base.

IBM's plan includes the creation of a Power Portal site where potential licensees can download a free Power Architecture Pack that includes software for designing a custom POWER chip. The idea is that chip designers can incorporate their own intellectual property (IP) and use it to enhance the POWER core in order to produce a chip--manufactured either by IBM or a third party--that's specifically tailored to their needs. Ultimately, IBM hopes to create a new market for

POWER derivatives of different types, used in everything from servers to game consoles to PDAs to routers, produced by licensees from across the industry.

Deja Vu

This model looks an awful lot like the DEC model of the 1980's. DEC was the premiere minicomputer vendor and it sold its hardware under its name or on an OEM / board level basis to companies who added a case, disk drives, software etc. and created a specific type of application machine. The hardware and software was bundled together as one and sold as a complete product. One of the examples I recall of a very successful DEC-based product was Dynix, which still is the leading university library system.

Just as Dynix would buy pieces they needed from DEC and then mold the box into a library system, the Power.org consortium provides a similar capability built around the IBM Power chip. The chip takes on its ultimate personality with the completed design from the Power client. IBM receives intellectual property revenue and the client receives a very powerful base chip design upon which to build a specific product, such as a cell phone. The difference between the old DEC approach and this is that the resulting product appears to be a hardware device with embedded software.

Analysts immediately began making comparisons between IBM's new open approach and Linux, but a better analogy might very well be Sun's Java. IBM will still control the PowerPC instruction set, at least insofar as the various POWER derivatives are concerned, and the POWER derivatives will all build on IBM's basic designs.

What's In It for IBM?

So why is IBM opening up the Power Architecture? What's in it for Big Blue? In today's computer world, it is clear that Microsoft has a virtual monopoly in software and its biggest threat is Linux. Intel has a virtual monopoly in hardware and its biggest threat appears to be AMD and there is no question that AMD is fighting Intel and the volleys are being returned in earnest. But, with the new independent IBM, does it not make sense that Open Power Architecture, with IBM in control of the core, could be far more threatening to Intel in all markets than AMD in the PC chip space.

There is no question that POWER is today's leading chip architecture. While Intel was concentrating for ten years on its Itanium project and it was offering token clock speed increases on its x86 chips; IBM was defining and building the finest processor chip family in the industry. Just like with Linux and Microsoft, IBM knows it cannot go head to head with Intel. The inner circle of Linux users and supporters knew that they could not go head-to-head with Microsoft. What Linux and FOSS (Free and open source software) has on its side is something Microsoft cannot buy-- openness. Without Openness, just as Microsoft gobbled up everybody in its path, including IBM on its way to monopoly status, Linux would have been another Microsoft statistic if it were not open.

All along in this book, we have been pointing out the real dynamics that are in play here in the chip business with IBM wide awake but pretending to sleep and Intel pretending to be awake but missing the daylight. IBM knows if it has a chance to win against Intel it cannot play the game by the old rules or Intel will continue with the lions' share of the market. The new rules of openness and collaboration will lead to growth in established markets much faster than having to have your own secret service fighting off Intel's men in black. Linux was not supposed to be about taking down Microsoft a few notches, but it sure seems to have worked out that way. Equally, the goal of IBM's PowerEverywhere effort does not seem to be to take over the desktop, but as we have seen with Microsoft, openness has a funny way of changing things.

Recognizing that IBM is a profit making organization, IBM is not going to just roll out its intellectual property for the taking but, it can certainly use the Sun Java model to control the very important elements and keep fully open the areas that would leverage the architecture for the open community. A good part of the GNU and Open Software libraries is not "free" per se, but it permits major innovation. For IBM it is a "have your cake and give it away too" notion.

IBM is marketing lots smarter in this chip business than it ever did in the mainframe era of the 60's. This is a new IBM that understands that the next step in computer technology is more important than clock speed on CPUs. Building out the technology that CPUs touch is the future. The POWER Everywhere effort from IBM will reach from PDA's to supercomputers, because openness allows accelerated growth that is normally kept down by proprietary intellectual property.

There is no question that IBM means business and if anything the new IBM is a patient being, waiting like a cat, yet moving forward, smiling, but not laughing, and always preparing its next move. Somehow, it seems at least publicly that IBM has been able to achieve a major market and development opportunity while still flying well under the Intel radar net.

Internal Use Only?

IBM touts that its prime purpose for its Power Architecture chips is in its own products such as the IBM eServer iSeries (formerly the AS/400 business computer) as well as its eServer pSeries (formerly the IBM RS/6000 Unix based computer). IBM says that it is engaged in the Power Architecture chip business to keep its mainline servers at the top of their game. Any use of the Power Architecture in non-IBM systems, such as Apple as in days past, is ancillary, not mainstream to IBM's goals for its server families.

Several years ago, this was not the case at IBM. Lou Gerstner, the consummate salesman believed that IBM should sell everything it had on the truck, even if it did not help IBM's traditional server businesses. So, he created two divisions for Power Architecture – the one division catered to IBM's own server families – basically the eServer iSeries (AS/400) and the eServer pSeries (RS/6000) lines of computers, while the Microelectronics Division solicited business for the use of IBM's chips from other companies.

POWER, PowerPC Development

Each division controlled its own destiny and during this time period, certain chip design decisions that saved miniscule amounts of dollars were made in one division to optimize that division's product for its intended customers even though those decisions would preclude the use of the new chip for other IBM purposes. During this time IBM engaged in classic sub-optimization. While each division's goals were optimized, the IBM Corporation's goals were sub-optimized.

In one case, for example, a design decision was made on the PowerPC that precluded the chip to be able to run the company's highly capable OS/400 operating system. Though many non technical people have never heard of OS/400 (renamed i5/OS in 2004), it is a major business operating system within IBM and the ability to run OS/400 could have meant that machines the size of Macintosh computers would be able to use IBM's most productive business operating system.

Recognizing that this was creating inefficiencies within the company, IBM changed its structure. Big Blue consolidated all of its Power Architecture work in one division. In this regard, IBM's server divisions now have the final say on which designs go into which iteration of the Power Architecture chips. Even with the influence of IBM's internal servers providing the guidelines for base functions, IBM's microelectronics unit has plenty of wiggle room to create some phenomenal designs using the Power architecture features as the base.

Game Consoles 'R Us!

As introduced in Chapter 10, for example, IBM has a major partnership underway with Sony and Toshiba for a phenomenal new chip called the Cell processor. This chip has a PowerPC as its base processor plus it has eight other processors to help Sony produce real smooth video as the computer generates images to show the game action. In the not-too-distant future, this chip is scheduled to be used in the next generation Sony PlayStation 3.

IBM is the unquestioned leader in game consoles. In 2003, in an unexpected move, Microsoft selected IBM's POWER technology fir its coming Xbox 360. IBM stole Intel's customer. Microsoft had been using Intel chips. The Redmond giant in fact is ready to use three next generation dual core POWER processors, perhaps POWER6 chips, in the new unit. Microsoft saw what Sony was doing with its Cell and signed up with IBM as a Power Architecture customer for a reason that can be explained in one word, "Sony." Microsoft was not interested in being left behind, regardless of its Intel relationship, and Power Architecture is the only way to go if you need chip power.

The other game console machine of the big three is the Nintendo. Nintendo has been an IBM chip customer for a long time so it wasn't big news that Nintendo decided to use IBM chips in its new Nintendo Revolution game console. IBM is not at liberty to say what chip will be used in the Revolution but Nintendo suggests that the chip is code-named Broadway. It is designed to co-reside with a new graphics chip designed by ATI and both chips, or sets of chips will drive the new Nintendo Revolution console when it hits the marketplace. Speculation is that the chip is in the POWER family and may very well be a special version ASIC chip.

It sure looks like IBM is set out to be the # 1 player in the game console and home processor marketplace. Since in a few years, that business will compete head-on with the PC business, it follows that, like it or not, IBM is definitely an Intel competitor.

No More Apples

With regard to the Power Architecture, IBM is in its fifth generation with the POWER5 iteration of this technology and Power6 just around the corner. So, IBM is really not a newcomer to the chip game. Until recently Apple was a big IBM customer but Apple has chosen to go to Intel. I am not inclined to think that Apple knows more than Microsoft, Sony or Nintendo, so I would expect that in 2006, Mac fans will be giving the Cupertino Company a hard time for having made that decision. There's more on Apple in this chapter.

PowerPC Architecture

In about 1990, the first 32-bit PowerPC chips were built in Austin Texas for use in the RISC System / 6000. The version of the PowerPC chip needed for Apple's Macintosh running Mac OS 8 etc. and the version needed for IBM's RS/6000 running AIX (IBM's Unix version) and the versions used for embedded processors did not need all of the computer science bells and whistles that were necessary to support the Amazon architecture (Page 188). So, IBM built lesser capable versions of the chip to accommodate these needs.

When the PowerPC project was initially brewing, IBM had not yet named its original POWER effort. So, retroactively, the company dubbed its first generation POWER technology as the POWER1. This gave a name to what everybody knew as the chip used in the 32-bit IBM RS/6000 machine when it was introduced in 1990. POWER then is the name of a RISC CPU architecture designed by IBM. This retro acronym stands for Performance Optimization with Enhanced RISC. IBM had called its RS/6000 models POWERstations so the retro moniker of "POWER" fits in nicely with past names.

Chip Name Confusion

Despite IBM's efforts with a retro name, the term PowerPC chip is often confused with the POWER chips used in IBM's internal processors. This is not always a correct assumption. Like the POWER architecture as it was called in the 1990's, the PowerPC is all about the design of a RISC microprocessor architecture and, as you might expect, it is based on IBM's original POWER architecture. The PowerPC moniker goes back to the 1991 Apple-IBM-Motorola alliance, known as AIM. From 1994 through 2005, chips derived from the original PowerPC architecture found their most success in the personal computer market in Apple's Power Macintosh.

The PowerPC and POWER architectures have remained close enough over the years that the same programs and operating systems will run on both if certain rules are followed.

The POWER series microprocessors continue to be used as the main CPU in many of IBM's servers, minicomputers, workstations, and even supercomputers. As noted previously, in today's world, where partnerships among semiconductor firms is becoming the rule, rather than the exception, IBM encourages other developers and manufacturers to use the POWER architecture or any other derivative of it through the Power.org (http://www.Power.org/) community that has recently been formed.

PowerPC History

It is a safe bet in today's semiconductor world of diverse chip designs that nothing comes from nothing. So, also with the Power architecture. The POWER design comes almost directly from the earlier IBM 801 CPU, widely considered to be the first true RISC chip design. IBM's 801 was a research project that came about because Big Blue wanted to build a box to test the theories of an IBM computer research engineer named Dr. John Cocke. Dr. Cocke worked at the Yorktown, New York facility. He had originated the RISC concept in 1974. Yes, Virginia, IBM invented RISC technology. Dr. Cocke proved that about 20% of the instructions in a computer did 80% of the work. The first modern RISC machine therefore was the 801 minicomputer built by IBM, in 1975 as a pure research project.

This was long before any commercial vendor even thought of building a RISC box. It was not until 1987 that Sun released its first SPARC system that RISC became a popular notion. Ironically, it was Sun and not IBM that made a killing on John Cocke's invention. IBM chose not to market the 801 and the company showed little visible interest in RISC technology because the company was still riding high on its prevailing CISC technology. IBM did not even publish anything about its work with RISC until 1982.

In 1982, there were several other research projects underway regarding RISC, one was the work of a research group at Berkeley led by a University professor, David Patterson, who, by the way is the person who named the technology "RISC." His project was known as RISC I and it led to an even better RISC II project. To be correct, Sun's SPARC series is actually modeled after the RISCII project and not the IBM 801, but Cocke had invented RISC seven years earlier. Another project at Stanford lead by John Hennesy started about the same time. This lead directly to the MIPS processors (www.mips.com) that are still sold today.

IBM RT/PC

IBM eventually got a project going in the mid 1980's from its 801 work. They called this the RT/PC. The best that I can say about the

RT/PC is that it used the Unix operating system, which was a departure from IBM's former stance on the widely used OS. Other than that history shows that the RT was a marketing dud. IBM's sales force had no idea how to sell it. Moreover, compared with other microprocessor driven systems of the day, it was a poor-performer.

In the mid 1980's, having learned the hard way that you cannot chinch on the design of a chip if you want it to be powerful, IBM began its infamous America Project. IBM's earnings and cash position were in good shape at this time so the company felt that it could bounce back from what it had learned about the RT/PC's poor design and create a design that was both revolutionary and spectacular. IBM's objectives for America were quite lofty. In no uncertain terms, IBM's intentions were to design the most powerful CPU on the market.

They were interested primarily in fixing two problems in the 801 design that carried over to the RT/PC:

- ✓ 801 required all instructions to complete in one clock cycle, which eliminated floating point instructions
- ✓ Although the decoder was pipelined as a side effect of these single-cycle operations, they didn't use superscalar effects

IBM had to solve the floating point problem first to be successful with its design. When IBM is challenged and it has the resources to apply to a technology problem, the problem normally gets solved. To solve this one, IBM used some reasonably new floating point algorithms developed in the early 1980s that supported 64-bit double-precision multiplies and divides in a single cycle. In other words, they solved the problem in a big way. There was no chinching. IBM also added the hardware to support the algorithms.

The Floating Point Unit (FPU) part of the hardware CPU design was built separate from the instruction decoder and integer parts. This allowed the decoder to send instructions to both the FPU and the Arithmetic/Logic Unit (ALU for integer calculations) at the same time. IBM also designed an innovative approach to getting parallelism in the instruction cycle. It complemented these two hardware pieces with an instruction decoder, which could be fetching one instruction, decoding another, and sending one to the ALU and FPU at the same time. So, the America project not only produced a state-of-the-art RISC machine, it resulted in one of the first superscalar CPU designs.

This first POWER chip set was 32-bit. It was well equipped for its 32-bit mission as IBM packed thirty-two 32-bit integer registers and another thirty-two 64-bit floating point registers on the chips, each in its own unit. The branch unit was also fine-tuned to include a number of "private" registers for its own use, including the program counter. All of these facilities contributed to the CPU's high performance.

The POWER1 Chip Set

The first POWER1 CPUs as introduced in 1990, consisted of three chips; branch, integer and floating point. The one chip POWER processor was yet to come. These were wired together on a nice sized motherboard to produce a single system. As previously noted, IBM used its POWER1 primarily in the RS/6000 series of workstations.

It was shortly after this time in history that IBM realized that it would need a single-chip microprocessor. To do so, because of the density of chip real estate at the time, the engineers would have to eliminate some POWER processor instructions so that the RS/6000 line could scale from lower-end to higher-end machines. Not long after this work, IBM realized that a single-chip POWER microprocessor could potentially become a high-volume microprocessor in many areas of the industry.

IBM and Apple

IBM approached Apple with the goal of collaborating on the development of a family of single-chip microprocessors based on the POWER architecture. Since Apple was one of Motorola's largest customers of desktop-capable microprocessors, Apple wanted Motorola to join in the talks. Apple also felt that Motorola had more extensive experience with manufacturing high-volume microprocessors than IBM and at a minimum; Apple wanted Motorola to serve as a second source for the microprocessors. This three-way collaboration quickly became known as AIM alliance, for Apple, IBM, and Motorola.

To Motorola, Apple, and IBM, the PowerPC consortium known as AIM (Apple IBM, Motorola) looked like an unbelievable deal. It allowed Motorola for example, to sell a widely tested and powerful RISC CPU for little design cash on their part. They also maintained ties with an important customer, Apple, and the deal seemed to offer the possibility of adding another customer. Motorola believed that IBM might very well buy smaller versions of the PowerPC from them instead of making its own.

At this point Motorola, however, already had its own RISC design in the form of the 88000 which was already doing poorly in the market. But, the 88000 was already in production. Adding to the negatives for Motorola, Data General was already shipping 88000 machines and Apple already had its 88000 prototype machines running.

This seemed like such the right deal that all three companies tried to figure out how to make it work for them all. If the new PowerPC single-chip solution could be made bus-comparable at a hardware level with the 88000, that would allow both Apple and Motorola to bring machines to market much faster since they would not have to redesign their board architecture. And so it was.

PowerPC is Born

The result of these various requirements was the PowerPC (Performance Computing) specification. Everyone seems to have won:

- ✓ IBM got the single-chip CPU they were looking for, largely for free
- ✓ Apple got to use one of the most powerful RISC CPU's on the market, and massive press buzz due to IBM's name
- ✓ Motorola got an up-to-date RISC chip for free, and help with design methodology from IBM

The first single-chip implementation of the PowerPC design was called the MPC601. It was a hybrid of the POWER1 and PowerPC specifications that had been released in 1992. Apple continued work on a new line of Macintosh computers based on the chip, and eventually released them as the 601-based Power Macintosh on March 14, 1994.

IBM Had PowerPC Workstations Ready for Windows

In a similar time frame, IBM's OS/2 had not quite died but IBM was getting closer to sending it to pasture. The word leaked out of IBM that the company had a full line of PowerPC based desktop units already built and ready to ship. However, the operating system, which IBM had intended to run on these desktops—Microsoft Windows NT—was not complete by early 1993, as IBM was ready to ship the units.

At the same time, IBM began to believe that it was getting double-crossed by Microsoft in a number of PC related operating system areas. So, this was about all IBM could handle. IBM decided that it would rewrite OS/2, its x86 based PC desktop operating system, for the PowerPC. It reportedly took IBM two years to rewrite OS/2 for PowerPC, but by the time the OS was ready, the market for OS/2 was diminished, thus the market for OS/2 on PowerPC had basically dried up.

IBM did ship some of these PowerPC 601 units, which were code named *Sandalbow*, based on the PowerPC 601 CPU, as RS/6000 models. Apple of course did not have a PowerPC operating system either but the company took a different route. They rewrote what they needed for performance purposes in their Mac OS operating system for its first release, and they also wrote a Motorola 68000 emulator which could run the rest of the Mac OS. Additionally, the emulator ran the Motorola 68000 library of programs. So, Apple was able to get its Mac OS out the door lots faster than IBM and OS/2.

As better PowerPC Chips were released, Apple rewrote the remaining parts of its OS and produced a pure version that ran native on the PowerPC models 603 and 604

The first 64-bit implementation of PowerPC was something called the 620, but this chip does not appear to have had much use since Apple didn't want to buy it. With its larger die area, it was too expensive to be sold in the embedded chip market so it appears that IBM had to swallow these efforts,, which it performed on Apple's behalf. Among other things, the chip was later and slower than promised, and, since Apple was not paying for development, IBM used its own POWER3 design instead, offering no 64-bit "small" solution. IBM finally announced the smaller, PowerPC 970 in 2002. The 970 is a 64-bit processor derived from the POWER4 server processor. To create this version, IBM modified the POWER4 core to be backwards-compatible with 32-bit PowerPC processors. Additionally, a vector unit similar to the AltiVec extensions in Motorola's version was added for full Mac compatibility.

Storms Brewing

All would not stay rosy in Denver. Though there were lots of good things about the alliance, it did not make much money for IBM or Motorola. Consequently, Apple's desire to move its line forward with new chips was not always easy for IBM or Motorola to accomplish. From Apple's perspective the same manufacturing issues began plaguing the AIM alliance in much the same way it did when Motorola single sourced Apple with its chips. Apple suffered from consistently pushed back deployments of new processors.

Since Motorola was Apple's prime vendor, they were the first to create problems for Apple in the 1990s with the Apple Macintosh G3 and G4 processors. Motorola's Freescale Division made Apple's PowerPC processors and after awhile, they decided they did not want to be in that business. Apparently Apple was quite demanding but their demands did not include any extra payment for the work necessary for accommodation.

So, with Motorola exiting the PowerPC game, IBM was forced to sole source and take over for Motorola. However, since Big Blue was even less agile in the small end as Freescale, the company also had problems meeting Apple's expectations with its 64-bit G5 processor in 2003 and its mobile processors. Apple was not happy with IBM and it created some bad press for Big Blue about its POWER technology.

In 2004, after consolidating its POWER line of processors, IBM admittedly was most interested in using its microprocessors for its own server use. Those in IBM who seem to know suggest that with Apple's low volume (2% of PCs), the company was really not making a profit on its Apple relationship.

In June 2005 Apple announced that it would no longer use IBM's PowerPC processors in its Apple Macintosh computers. Industry observers speculate again that this has more to do with not having most favored nation status within IBM than any technology issue with the IBM chips.

IBM Sells its PC Business

At about the same time, IBM began to exit the personal computer market completely by selling its entire consumer line of PC products (which all used Intel processors) to a Chinese company named Lenovo. Some think that IBM is planning to introduce new PowerPC systems from Lenova that leave all other desktop and Laptop systems in the dust.

I don't really see a relationship between the PC sell-off and Apple. IBM's first POWER chip priorities have always been for its own servers (eServer iSeries and eServer pSeries). Perhaps Apple did not have as much input to the chip design as they would have liked. Another line of thought on Apple is that IBM began to focus its attention on its chip designs for PowerPC towards game machines such as Sony's PlayStation 3 and Microsoft's Xbox 360 and Nintendo's Revolution. Though IBM is not talking, there is considerable reason that this theory may be right. With all three major league game console vendors, Sony, Nintendo, and Microsoft using IBM chips, the volume of chips coming to game consoles in the near future is certainly worth IBM's attention.

With all three game console vendors under contract, including Microsoft, a long-time Intel partner, IBM has good reason to believe that these companies, looking for the best processor possible for their performance needs, obviously see something special in IBM's abilities. Logic suggests that there cannot be a good technical reason for Apple to abandon this superior technology to form a marriage with Intel to use its new x86 dual core processors for next generation Apple units. On the other hand, perhaps Apple will select an Itanium derivative, and if this is the case, with Itanium's market acceptance not meeting anyone's expectations, then all bets are off.

Apple has been quoted as citing the performance limitations of the PowerPC chip for future personal computer hardware specifically related to heat generation and energy usage going forward, as well as the inability of IBM to move the 970 (PowerPC G5) processor to the 3 GHz range. However, it appears that the real reason is that IBM is not moving fast enough for Apple. As a sidebar, it also appears that IBM R&D's chore of pleasing Apple comes out of IBM's pocket as Apple apparently was not contributing to the major R&D load that it was creating.

Intel Hates Microsoft?

Though the public perception that something was wrong in IBM is not true, this was a definite public relations problem for IBM. The Apple decision appeared to be made in part by IBM's inability to match Intel and other competitors in terms of speed and architecture improvements. This logic, however, does not stand the reasonability test. A more probably reason is that Intel really hates Microsoft. With Apple in the stable, Intel has an opportunity to beat Microsoft in the desktop wars by bolstering Apple's presence. That makes a lot of sense and perhaps that really means that in thse Chip Wars, we should count Microsoft in. It wouldn't be like the software giant from Redmond to stay out of a good fight.

Only Nosterdamus knows for sure but from my eyes, Apple pulled out just at the wrong time, not just at the right time. Steven Jobs obviously did not get what he wanted so he took his fruit and went home. Cupertino and Santa Clara are much closer geographically than

Armonk, New York, and Redmond Washington. So, maybe it's OK that Steve went home.

For those not familiar with all of IBM's current chip ventures, the Apple and Freescale pullout may have appeared to leave the future of the PowerPC platform in much doubt. However, in addition to its own servers and to the game console industry, IBM is also making a great business out of embedded devices (ASICs) based on the Power Architecture. IBM today offers the best technology in this growing area. So maybe Apple's pullout is OK. In any case, the Power Architecture that IBM developed is still very much alive, and I predict Apple fans will rue the day that Apple chose Intel over IBM.

POWER History

Well, that's the PowerPC story. But, the PowerPC is not the chip that has driven IBM's internal servers all these years. Soon after POWER1 came POWER2. The single-chip POWER2 processor was much improved over the POWER1 and it lived the longest of all the POWER generations. It was released in 1993 and it was at the top of its game until 1998. In addition to all of the POWER1 goodies, the POWER2 added a second floating-point unit, 256 kB of cache and 128-bit floating point math capability.

The POWER3 processor followed in 1998, moving to a full 64-bit implementation, while it remained completely compatible with the POWER instruction set. This had been one of the goals of the PowerPC project and the POWER3 was the first of the IBM processors to take advantage of it. The design added a third ALU and a second instruction decoder, for a total of eight functional units. IBM's POWER processors had clearly become the leaders in parallel functions on a chip.

In 2001, The POWER4 model presented an industry first two complete CPU cores on a single one chip. Otherwise, the core design was very similar to the POWER3. Additionally, IBM added high-speed connections to up to three other pairs of POWER4 CPUs. The IBM design was such that the four chips can be placed together on a motherboard to produce an 8-CPU symmetric multiprocessing (SMP) building block. When processing requires high throughput instead of high code complexity, IBM built in facilities can turn off a pair of cores so that the remaining cores can have the entire bus and L3 cache to themselves. The POWER4 in its day was considered by many to be the most powerful CPU available.

To help IBM's friends at Apple, as noted above, the company introduced a single CPU core version of the POWER4 called the 970. It is employed in the newest generation of Apple desktop computers (that is, the G5).

Three years after the revolutionary POWER4 processor, in 2004 IBM introduced its POWER5 processor running at 1.9 GHz. Though this was a two-processor chip, even with one processor turned off the POWER5 posted the highest uniprocessor SPECfp score of any shipping chip. In 2004, IBM introduced its eServer p5 as the newest model of the former RS/6000 line and it introduced its eServer i5 as the newest model of its AS/400 line. The POWER5 powers these two eServers.

IBM became very stingy between POWER4 and POWER5 with chip real estate. The number of circuits in both chips is about the same. Yet, with the same number of circuits, IBM was able to introduce many improvements in the POWER5 over the POWER4 which include: a larger L2 cache, a memory controller on the chip, simultaneous multithreading, which appears to the operating system as multiple CPUs, advanced power management, dedicated single-tasking mode, hypervisor (virtualization technology), and eFuse (hardware re-routing around faults).

IBM was very proud of its work. Ravi Arimilli, IBM's chief microprocessor designer offered this comment: "The POWER5 chip is more of a midrange design that can drive up to the high end and then down to things like blades." IBM's eServers (i5 and P5) are now built with POWER5 processors that offer virtualization features: logical partitioning and micro partitioning. Up to ten partitions can be created and driven for each CPU.

IBM's CTO, Bernie Meyerson offered that the POWER5 is so rich in performance that jut one half of one core on the chip or ¼ of the chip's capabilities is more than enough to control a flight to the moon.

The rest of the specs on IBM's eServers, which were built as part of a holistic design (See Chapter 23), are so amazing that they would appear to a computer scientist to be untrue. IBM introduced the most powerful computer in the world at the time the POWER5 was introduced in 2004 – twice as fast as the mainframe at that time. The eServer i5's biggest 64-Way system can run 256 independent operating systems. Moreover, memory, CPU-Power and I/O can be dynamically moved between partitions. The POWER5+ is around the corner and IBM's schedule has its POWER6 chip slotted for next year. One might ask how can there be more. It actually is amazing, and perhaps it is also Amazon.

Amazon Architecture

Sometimes it is actually seems too confusing for even IBM to differentiate PowerPC and POWER architecture. A you now know, the POWER architecture and the PowerPC are very close relatives and they lived and ruled the day in Apple Units and IBM's RS/6000 until about 1998. However, the 32-bit PowerPC and the 32-bit POWER chips that were successful in these ventures could not be used in IBM's most advanced server. They were too small and did not contain enough hardware facilities for IBM's own AS/400 box, which had been based on a 48-bit CISC hardware architecture. Moreover, there are some phenomenally advanced one-of-a-kind computer science features that had been implemented in the AS/400 that required special devices to be created on the chip for a new machine to be able to function as well as the to-be-replaced 48-bit AS/400 CISC unit.

So, in early 1991, a highly talented team of people from IBM development locations in Rochester, Minnesota, and Austin, Texas, got together with their counterparts from IBM Research locations in New York to define a new processor architecture. The intent was to create a new architecture that could be used for all future models of both the AS/400 and the RS/6000. IBM called that architecture Amazon.

A chip to support the AS/400 specifics would have to be big indeed. So, the Rochester engineers took the POWER design and modified it to work with 64-bits. In addition, they added extensions to the chip to support the advanced notions that were needed on AS/400. Together the design was acknowledged by all as Amazon.

Adding Missing Pieces to the Architecture

The RS/6000 POWERStation was a single-user workstation as is the Macintosh. Thus, the original POWER processor and its derivative, the PowerPC processor and their associated architectures were designed to be the engines for single-user workstations. Yet, the AS/400 and later models of RS/6000 supported by Unix (a multi-user operating system) needed more devices on the chip than were available to fully support multi-user. In order to use either of these Power Architectures in a multi-user, multi-application server, many new functions had to be added. For example, neither the original POWER nor the original PowerPC architectures had support for multiprocessors.

More than any other feature; the game buster for AS/400 was that the architecture did not support one of the important advanced features of the AS/400 known as single level storage. The advanced addressing mechanism of the AS/400 required that substantial innovation occur at the chip level to support this important feature. AS/400 could not use the addressing mechanism that was originally created for POWER and later adapted for PowerPC. As just about everything in the original RISC Power chip, its addressing mechanism assumed a Unix style of addressing where each Unix process has its own private address space. Single Level Storage assumes a shared address space for all processes and as such it inherently provides a data and program cache in RAM. To support single-level store in Amazon, therefore, IBM had to define a second complete addressing mechanism to exist on the same chip.

In addition to AS/400 specific features, some features from the original high-end RISC System/6000 POWER chip set needed to be rolled back into the generic Amazon architecture to support high-performance technical computing. Many of the features that had been part of the original POWER architecture were not included in the more general PowerPC architecture to save chip space. What really happened was that by trying to accommodate all of the unique features of high powered technical computing as in high end RS/6000 machines, and the advanced architecture features of the AS/400, in the end, IBM had designed another new processor architecture. So, for future compatibility, Amazon would include the full PowerPC architecture, both 32 and 64-bit versions, along with all of the extensions needed for 64-bit implementations of the AS/400 and the RS/6000.

The AS/400 Advanced System/36

Way back in 1994, the first run of POWER designed chips was used for the re-issuance of an old IBM small business computer known as the System/36. IBM had discontinued the System/36 six years earlier but brought it back by popular demand using the newest POWER chip. At the time, Big Blue had not yet perfected the 64-bit RISC chips that would be used in its AS/400 models but it was able to put the full S/36 instruction set on the POWER chip for its new S/36 units..

IBM introduced the new machine as the AS/400 Advanced System/36, a PowerPC-based version of the AS/400 that natively ran the System/36 SSP operating system and its applications. It was not until 1995, a year or so later that IBM was able to load the full Amazon architecture on the POWER chip as the company changed its AS/400 machine from 48-bit CISC to 64-bit RISC

The Rochester-designed 64-bit RS/6000 chips were initially said to use RS-64 architecture, just as the AS/400 used the Amazon approach. However, as noted above, two different chips did not sound like a good idea to IBM. So, all of Amazon facilities plus the RS-64 instructions needed for the RS/6000 were put on the new chip. IBM did not build two chips, just one. Both chips were the same. Amazon was an all-inclusive architecture.

To the Victor, the Spoils

For awhile in the early 1990's, there were two IBM labs designing 64-bit processors. Rochester's AS/400 team was doing a good job of getting all it needed on its AS/400 chip and Austin's RS/6000 team was doing its thing. Unfortunately for Austin, its64-bit PowerPC effort for RS/6000 failed miserably. The Austin Lab actually had to rely on its AIX partner Bull to create its 32-bit PowerPC SMP designs. These machines got stomped on by rivals Sun Microsystems and Hewlett-Packard. IBM Austin, where the RS/6000 and pSeries originally hailed from, always got a lot more credit from the press than it deserved for the Power line of servers. IBM Rochester was the leader in most of the major chip design innovations.

To this end, to reward the successful and to be able to avoid the unsuccessful, IBM moved its major chip design efforts for internal servers to Rochester and it also moved its manufacturing line to this IBM lab in northern Minnesota . Because the impetus in design was to create a unified chip, with all facilities for all "Power Architectures," IBM also gave Rochester control over the original PowerPC architecture. All chips, once designed were fabricated in East Fishkill, NY and sliced in Burlington, Vermont.

When the 64-bit RS/6000 system products came off the Rochester line, they were almost identical to the AS/400s. They both have used the same chip now for five years. The AS/400 uses the Amazon architecture on the chip and the RS/6000 uses the 64-bit PowerPC architecture on the chip – but it is the same chip. The chip has what is needed for both. When IBM introduced POWER4, the Power architectural merge was completed.

The Operating System Effect on Chip Personality

Operating systems provide the personality to all systems. Different operating systems use different processor architectures. For example, OS/400 uses the Amazon architecture with the addressing structure that is designed to support single-level store. AIX originally used the 32-bit PowerPC architecture but it has been enhanced to now use the 64-bit PowerPC architecture. Linux originally used the 32-bit PowerPC architecture, but with the introduction of the 64-bit Linux kernel, it too can use the 64-bit PowerPC architecture. Applications also use the different processor architectures. OS/400 applications use the 64-bit Amazon architecture. AIX and Linux applications use either 32- or 64-bit PowerPC, depending on the processor target when the application was compiled.

All these capabilities are on the same chip. One might ask "How does the processor know what architecture to use?" The answer takes just one sentence. Settings in the internal-processor control registers tell the hardware which architecture to use. These control settings are carried along with each process in the system and are loaded into the right control registers when the process becomes active. Thus, from POWER4 on, the chip has the ability to switch architectures as it moves from executing one process to another. Depending on the bit settings in the control registers, the processor can interpret instructions that look very much alike, very differently.

How Difficult Would It Be?

So, how difficult would it be to add the mainframe architecture to the POWER4 style architecture? I would say it is on its way. So, how difficult would it be to add the PC's x86 architecture to the POWER4? As noted in Chapter 10, IBM already did that in 1993 and took it away. Just like IBM added back in the removed POWER instructions from PowerPC and it added the necessary instructions for AS/400 and it supports both 32-bit and 64-bit computing on the same, one thing is for sure, if it thinks there is a market, IBM will build it, and from past technology accomplishments, it will work.

Over the years, from the original in 1990, the POWER RISC (reduced instruction set computing) architecture has gone through a number of iterations. In the late 1990's for example, Rochester IBM, with all its successes got very confident in its abilities. The Lab began to design the "Star" series of processors: These were to be delivered one each year to keep the performance of the AS/400 at peak. Northstar, Pulsar, IStar, and SStar were the specific code names. Because there were lots of different computers in IBM that would depend on the next "Star" POWER chip, IBM had to add more and more instructions on the chip.

Figure 15-1 64-bit Amazon / Power Architecture over Time

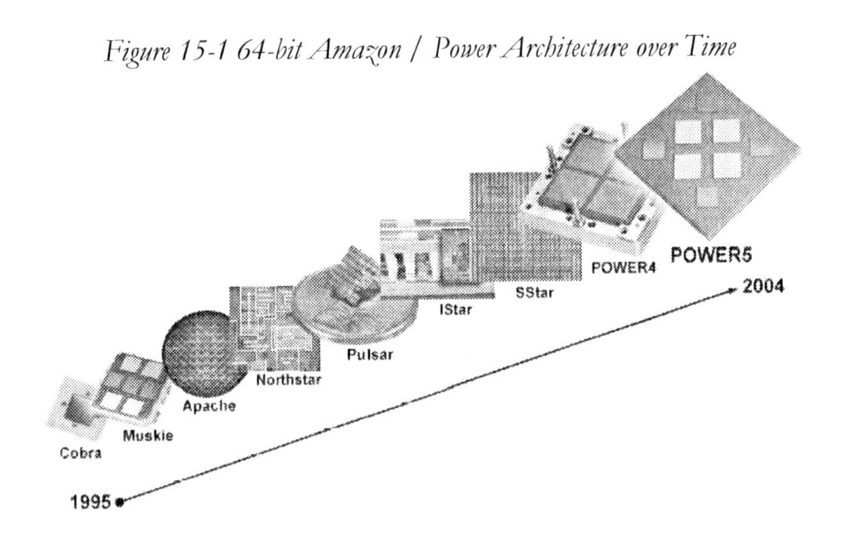

Complex RISC Computing

When the Star series came out, it implemented three architectures on the one chip. These three architectures are: 64-bit Amazon, 64-bit PowerPC, and 32-bit PowerPC. Of course, these three architectures are closely related to one another. The 32-bit PowerPC architecture is a subset of 64-bit PowerPC, which in turn is a subset of 64-bit

Amazon. Since the AS/400 was originally implemented in 64-bits in 1995 on the Cobra processor, there never was a 32-bit Amazon. Amazon exists only as a 64-bit architecture.

Because of all the instructions on the POWER chip today, there are those in the industry who suggest that the current POWER chip is no longer RISC, but is really a CISC (complex instruction set computing) based chip. Others say that there is no real difference any more between CISC (e.g. Intel x86) and RISC (e.g. POWER) since all modern chips are so complex and have so many instructions, the definition no longer holds. IBM continues to call its POWER chips RISC, however, because that was their origin, and the company refutes the experts' opinions and instead has the Power Architecture slotted as a "modern" RISC processor.

RISC was to be the great simplifier and when it was first introduced it was substantially simpler than CISC. The System/370 instruction set is typical of CISC architecture. It has 183 instructions defined. An early non-power RISC implementation (MIPS) contained less than 40 instructions, but those days are long gone. Today's RISC based Power Architecture has had enough instructions introduced to the chip over time to support all the variants of RISC in the POWER line, and thus there are substantially more instructions implemented today in Power RISC than there were in System/370 CISC.

Let's look at the relationships of some of the Power Architectures over the years that IBM has been able to combine on the POWER chip. The original 32-bit PowerPC architecture started off with 187 instructions. The 64-bit PowerPC architecture added another 41 instructions for a total of 228 instructions. The original Amazon architecture added another 25 instructions to bring the total to 253 instructions. By the time hardware convergence between the AS/400 and RS/6000 had been achieved in 1997, the number of instructions on the chip had grown to 385. And the number is continuing to climb.

The number of instruction on the chip is continuously changing, because new instructions need to be added each time a new processor is developed that provides new functions. For example, new instructions were added over time to support Linux on the iSeries. Similarly, new instructions were added to the later processors in the

"Star" families to support partial processor logical partitioning (LPAR) functions. As a result, the later "Stars" in IBM's chip lineup had more instructions than the earlier "Stars." This also helps to explain why some new software functions run only on the newer microprocessors.

Looking at IBM's chip accomplishments, only Nosterdamus knows for sure, but from my eyes, Apple pulled out just at the wrong time, not just at the right time. Steven Jobs obviously did not get what he wanted so he took his fruit and went home.

The IBM Cell – Supercomputer on a Chip

The big star in IBM's POWER lineup may very well be its new Cell processor, which is unique in the industry and it was designed as a joint effort of Sony, Toshiba, and IBM. In addition to being a nine processor implementation, something that Intel is not close to doing yet, this is also a consortia effort among several companies. Unlike this IBM venture, Intel typically goes it alone.

Figure 15-2 IBM Cell Processor (Power Plus 8 SPEs)

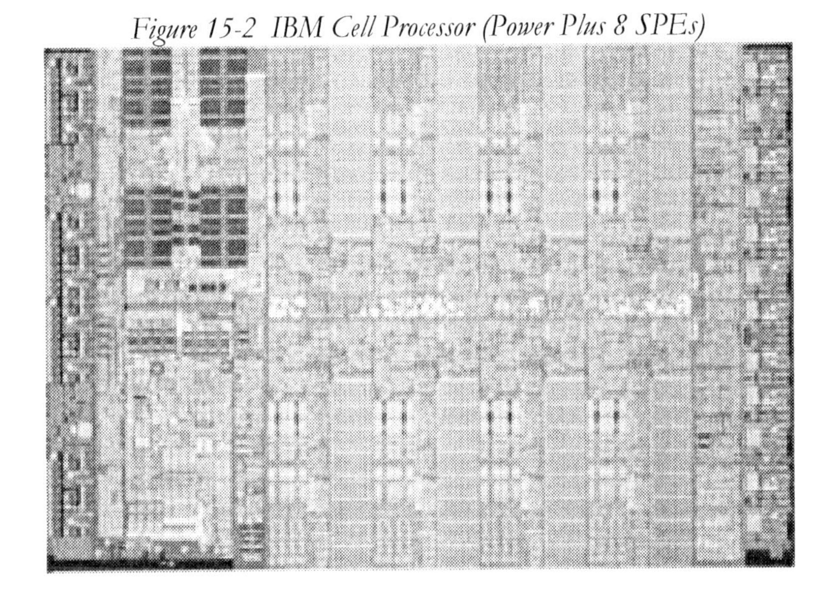

Intel has traditionally tried to win chip business using its standard chip set. However, the x86 processors have not typically been too affordable and that is how IBM first got Nintendo's business. Though Intel does have its share of the ASIC / embedded systems market, and it has been the Microsoft Xbox chip supplier since its inception, it does not have any chips in its arsenal designed specifically for the game console industry.

Meanwhile, in Rochester, Minnesota and Austin Texas, IBM continues to expand its POWER work with an even newer chip design for the game console industry. The Cell Processor has already been introduced in Chapter 10. In this chapter, we discuss some very interesting specifics about this "supercomputer on a chip."

How about we start with some good but true hype: When coupled together with other chips, the Cell processor, which is constructed based on a supercomputing model, can reach speeds that are 1000 times faster than current technology. In its one-chip state, Cell delivers power that is up to ten times faster than any other chip being made today. Because of its anticipated use in the new PlayStation 3, Sony is predicting that the new unit with just one Cell chip will run 1000 times faster than the PlayStation II. The following extended snippet from PCstats gives a nice industry perspective on this latest and perhaps greatest IBM POWER chip.

http://www.pcstats.com/articleview.cfm?articleID=1727
IBM's CELL Processor: Preview to Greatness?
February 11, 2005

"IBM, in cooperation with fellow industry giants Sony and Toshiba, recently announced a new processor, called the Cell. While this in itself is not really cause for much celebration (except perhaps for the odd bedfellows involved in the project) the new multi-core chip might well prove to be something special. For one thing, it's going to be at the heart of Sony's upcoming PlayStation 3 console, which we have a feeling may be slightly popular…

The Cell processor is vastly different from conventional processors inside. This tiny chip contains a powerful 64-bit dual-threaded IBM PowerPC core but also eight proprietary 'Synergistic Processing Elements' (SPEs), essentially eight more highly specialized mini-computers on the same die.

It's these SPEs that make the Cell architecture special, as you might guess. IBM describes the product as a 'System on a Chip.' Like IBM's Power5 processors, multi-processing is build right into the die.
...
PCstats is going to take a quick look at what's currently known about the architecture of the Cell processor and its potential as a rival to today's x86-based 32 and 64-bit processors.

The basics: Cell Biology

As we mentioned, the prototype Cell processor is composed of a single 64-bit RISC PowerPC processor and eight SPE 32-bit units. These are bound together by a fast internal bus, the Elemental Interface Bus (EIB). A built-in dual channel memory controller is included, and connects to a current maximum of 256MB of extremely fast Rambus XDR memory.

Communication with the rest of the system is provided by the FlexIO bus. This interface also allows high speed, chip-to-chip communication between different Cell processors, either inside or outside the same computer system.

The prototype Cell processor ran at 4GHz, and according to IBM, is capable of a theoretical maximum of 256Gflops, thus placing it instantly at the forefront of potential multi-chip supercomputer designs of the future. The chip is built on a 90nm process and contains 234 million transistors....

But just as important, the extra computing power could help transform the game console into a home's primary source for delivering music, movies and Internet-based entertainment -- all areas where Sony already has interests.

More than 400 engineers, primarily at IBM's semiconductor design center in Austin, have worked on the project since the three companies started collaborating on it in March 2001.
In all, the companies have spent more than US$2 billion on the design and to retrofit chip factories in Fishkill, New York, and Japan that are scheduled to start producing the chips later this year.

The new Cell chips can support virtually every type of operating system, IBM claims. They also can be virtually linked to other Cell chips, increasing their performance potential even more.

The new chips are also unique in that they can have up to nine "cores," or processing units, allowing them to handle up to 10 different software operations at the same time.

In contrast, makers of personal computer chips are only starting to push into multi-core processing, a technology that IBM pioneered."

Nine Processors in Cell

The Cell is nothing short of a supercomputer on a chip. As noted previously, it is expected to be 10 times faster than current technology with its 8 side-bar high-performance processors that are designed to make graphics and video flow in real time.

It is interesting that IBM describes the Cell chip as a system (holistic design) as opposed to a microprocessor. In many ways it has the architecture of a supercomputer complex on one chip. IBM's iSeries and AS/400 boxes, which also are based on the company's Power Architecture for years have been equipped with separate input/output processors to offload the main processor from having to control all of the events on the system. As a system, therefore, these processors all talk together in harmony working on the problem at hand. Unix boxes are not designed like this so many heavy techs with a Unix background would not immediately relate to a main processor working with satellite processors.

The Cell processor is quite a machine and IBM intends to use it for more than game consoles. While Sony wants to use the Cell's power to show off virtual firepower capabilities, a company called Mercury Computer Systems is about to use the Cell for some real-life military applications. The company's technology is used in a wide variety of "defense and aerospace" applications. In military reconnaissance and surveillance platforms the company's systems process real-time radar, sonar, and signals intelligence data. IBM is partnering with Mercury in this venture. The message here is that the Cell processor is being noticed in the industry for its extreme performance capabilities, not just for its entertainment bias.

There is nothing in the Intel product line, the Intel pipeline, the Intel design lab, or the Intel dream lab that compares with the Cell processor. It's one of a kind, and it's a Power Chip from IBM.

Rochester Minnesota: IBM's Cell Processor and Supercomputing Lab

It is very understandable that IBM has selected its Rochester Lab personnel to work on the Cell initiative along with the Austin Lab since Rochester was the lab that created the AS/400 line and it is now home for the newest AS/400 product that IBM calls the eServer i5. Rochester was very involved in the movement of the POWER line of processors to 64-bit RISC. Moreover, Dr. Frank Soltis, IBM's iSeries Chief Scientist and the guru behind much of the design of the Amazon / Power Architecture is based in the IBM Rochester Lab. With all Rochester systems as well as IBM's supercomputing initiatives being based in Rochester, it is no wonder that a Power Architecture chip is the major engine for the Cell processor.

Industry analysts are beginning to understand that IBM has more than a passing interest in supercomputing. If we were to go back in history a few years, we would find that IBM management was once obsessed with supercomputing. Until recently, however, IBM just was not good at it. In the past, the tools for building powerful commercial processors and building one-at-a-time supercomputers were just too different for IBM to be good at both. That model has changed big-time and IBM is now uniquely positioned to be the best at both.

IBM and Supercomputing

IBM's interest in supercomputing goes way back to the Watson days when both senior and junior Watsons were continually embarrassed by less conservative companies, such as Cray, who were able to exploit the latest technology in their scientific wares. Because IBM was mostly concerned with its commercial processing capabilities, meaning its machines had to be affordable, its research and its development and its implementations were always a hair short of the processor power needed for the Watsons to gain any major pleasure in IBM's commercial prowess.

Those days are gone. With RISC and the 64-bit dual core POWER processors, and the 9 processor Cell units, finally what is good for the goose really is also good for the gander. By performing superbly in a multi-chip, multi-processor role in supercomputing, IBM is learning far more than any other company how to package these processors on silicon as they are miniaturized enough to control game boxes, home security systems, automobiles and lots more. The Cell processor itself is a reflection of all IBM has learned in its supercomputing research projects.

Supercomputing Power at IBM

In Supercomputing for example, as noted in Chapter 24, IBM has an effort underway called BlueGene/L A trip to www.top500.com, which lists the top 500 supercomputer sites in the world, will show that the top two sites and many other sites on the list are using IBM's BlueGene technology that is based on POWER processors.

IBM has the biggest supercomputer of all time in development. In just one of these supercomputers that is in process at its Lab in Rochester Minnesota, when compete, IBM will have ½ of the supercomputing power in the world in one machine. Talking about miniaturization and multi-core vs. dual-core processing, this project is deploying 32 POWER processors on each chip with over 1 million processors in total – all working together. IBM's Lab in Rochester Minnesota has always taken pride in its own style of getting things done. In other words, Rochester always thinks outside the box as an IBM Lab and never really has gotten into the prevailing MIT / Harvard, Cambridge computing models.

Major Design Challenge- Planning for Failure

At such a high number of processors (1 million), one of the design challenges for the development team was to deal with the reality of processor failure rates. The fact is that all electronic components fail. Nobody wants computers to fail ever but exceptional, not perfect reliability is all one can hope for.

With 1 million processors to keep alive at all times, the problem is exacerbated. If there is a one in a two million chance that a processor will go bad on a given day, and you have one processor, chances are yours will never fail. However, if your complex has a million processors, with today's technology, you can expect that one processor will fail every two days. Of course, nobody wants to have to repair a system every two days. So, a system with 1,000,000 processors would have to be built as self-healing. In other words, when a processor dies, the system itself would have to recognize it, isolate it, and call on a spare processor to take up the load while the failing unit is replaced– all without missing a calculation.

While Intel was making its living on stretching its PC power by increasing its single thread speed, as you can see with these supercomputing efforts, IBM was figuring out how to manage millions of computers at once in parallel. And, if one day all million or more of those little guys fit on a chip or two, IBM's supercomputer research will have paid off in practical uses for its leading edge processor line. That day is on its way and it is coming fast.

Not too many years ago a computer took several chips. Even the POWER1 took three chips to build. IBM has been investing its energy in developing big systems with many processor chips so that the chips can rapidly communicate with each other. As densities increase on chip real estate, IBM, especially with its Cell processor design has been able to take the concepts of driving multiple CPU chips in harmony, to the next level. IBM builds multiple processors on its chips and it is the only company with the supercomputer experience to know how to have them all talking at the same time.

And, though it may not be good for people to talk at the same time, to have nine computers as in Cell working on the same problem at the same time is called parallelism. It's like having nine Budweiser Clydesdales carrying the load instead of one tired old pack horse. Now, doesn't that make you thirsty?

Intel is not really engaging IBM in Supercomputing, though the company does have its own initiatives. The fact is that Intel is not in the system market; it is in both the circuit board and chip markets. Intel does not make servers. You may see Intel inside on lots of machines, but that machine also says, Dell, or Gateway or HP on it as the system manufacturer. Other than the AMD servers, IBM machines have IBM inside and IBM outside. It stands to reason that a company that can place one million processors in a supercomputing box and create a functional, usable system has a technology leg up on company's that simply can't.

Other POWER Goodies

Besides multiple architectures and a vast instruction set to support pSeries, iSeries, embedded processors, partial logical partitioning, and other advanced facilities, IBM is loading more and more 'power" on its dual core 64-bit POWER chips. In the remaining chapters, IBM's Power Architecture and IBM's many on-chip enhancements as well as AMD's leading edge technology are compared and contrasted with the Intel x86 and Itanium lines. In these chapters, we delve into things as mundane as research and as exciting as the mysteries on the chip that make one chip better than another. In the end, we offer our thoughts on the clear leader in the processor chip industry today.

Chapter 16: Intel: The Clock Speed Leader

Intel – The Last Three Years

Let's start the rest of this book by examining some of the braggadocio that exuded from the 2002 Intel Developer Forum (IDF). At the forum, the top dogs at Intel offered their thoughts and predictions on the significance of Moore's Law to chip manufacturing. The execs could not resist noting that "the law" was going to continue well into the future as the basis for Intel providing more and more processing power to users by merely increasing the density and clock speed of its famed "Intel Inside" chips.

Foremost among the presenters was Intel's Chief Technology Officer Patrick Gelsinger. In an address to more than 4,000 engineers and developers at the IDF, Gelsinger said that Moore's Law will drive rapid chip innovation for decades to come and that its influence will extend beyond digital devices to new areas such as wireless, optical and sensors.

"Our intention is to take the rapid pace of innovation and decreasing cost associated with Moore's Law and expand them to new areas that have yet to benefit from the kind of complex, highly-integrated silicon-based technologies we develop," said Gelsinger.

At the Feb. 25, 2002 session, then Intel CEO Craig Barrett also got his say: "As developing nations strive to build knowledge-based economies, technology is driving prosperity and productivity. Silicon technology and the integration of digital computing and communications are at the heart of underlying technology advancement and progress."

During his remarks to approximately 4,000 developers, engineers and other technical experts at the IDF, Barrett also praised Moore's Law and noted that both the "Law" and the Internet are not only alive and well, but are the driving forces behind fundamental economic growth. He said that Intel plans to be at the forefront of providing the computing and communications industries with the core technologies that will help provide the foundation for continued technology innovation and the growth of the Internet.

Barrett continued by saying that by moving forward with the company's aggressive technology and manufacturing programs, Intel was not far from seeing the day when the chip giant can put two billion transistors on a chip that operates at speeds of up to 30 GHz. He then summarized by saying: "These are the kinds of technologies that will help our customers build great new products."

Along with a number of other great pieces of braggadocio, Intel demonstrated a Pentium 4 running at 4GHZ. The company noted that within months, the first production version of 4GHz Pentium 4 chips would be rolling off Intel's fab lines.

A funny thing happened on the way to delivering on its promises. To take you through what actually happened, I have included a number of press clippings below. You will almost be able to feel Intel's embarrassment.

The following is a piece of an article from PC World in 2004, just about two years after Intel had boasted about how the implied corollaries of Moore's Law was about to continue to pay huge dividends in performance.

http://www.pcworld.com/news/article/0,aid,117176,00.asp
Intel Delays 4-GHz Pentium 4

"Setback is just the latest in a tough year for the chip maker." By Tom Krazit and Robert McMillan, IDG News Service
Friday, July 30, 2004

Intel has decided to push back the launch date for its 4-GHz Pentium 4 desktop processor to the first quarter of 2005, after reviewing its launch schedules and determining it would not be able to introduce the product in sufficient volume, a company spokesperson says."

As Intel was struggling in 4Q 2004 to keep its promises from almost three years earlier, the next snippet from PC World shows how the end of major performance increases at Intel *via ramping up clock speed* was more than likely near its end. Though Intel had assured the world in 2002 at its developer forum that 4GHz was just around the corner, near the end of 2004, with a modicum of humility, the company was forced to say that it couldn't do it. This snippet from PC World gets to the crux of the matter.

http://www.pcworld.com/news/article/0,aid,118424,00.asp
Intel's Pentium 4 to Top Out at 3.8 GHz

It's probably the last increase in this processor's clock speed.

Tom Krazit, IDG News Service

Monday, November 01, 2004

"In two weeks, Intel will increase the clock speed of its Pentium 4 processor for possibly the last time as the company heads into a new era for its desktop processor designs.

The Pentium 4 570 processor is a 3.8-GHz chip with 1MB of Level 2 cache. It will feature the fastest clock speed of any Pentium 4 processor for an indefinite period of time, and will lead Intel's mainstream desktop segment when it is released on November 15, an Intel spokesperson confirmed Monday. Intel's shift away from clock-speed frequency as a central design philosophy has been well documented this year. In May, the Santa Clara, California, company canceled plans for two high-frequency single-core processors in favor of an acceleration of dual-core designs due by the end of 2005. More recently, plans to release a 4-GHz Pentium 4 were nixed last month after Intel decided the effort required to reach that milestone would not be worth the expense."

Intel Great for PCs

There is no question that Intel's chips have been absolutely great for PCs and for single application servers. Intel will more than likely continue to do well in this arena where single applications on single servers and server farms are the norm. Certainly Dell, as noted in the following snippet from 2005, has affirmed that it will be using Intel chips way into the future.

Dell Chief Affirms Commitments to Intel, Blades, And Global Sourcing April 5, 2005

By Larry Greenemeier

InformationWeek

"Dell CEO Kevin Rollins had a number of things on his mind Tuesday as he addressed public-sector IT users and managers at the FOSE 2005 conference in Washington, D.C. Topics during his Q&A-style keynote ranged from the company's relationship with Intel to the impact of shrinking technology components on the data center to the company's perspective on call-center outsourcing.

Intel will play an important role in Dell's quest to deliver increasingly compact servers. 'As we increase density, power, and performance, there is an increase in heat generated,' Rollins said. One solution is Intel's multicore chip development. 'Multicore processor technology will allow us to mitigate the heat generated by systems,' he added.

... Intel's multicore processor design is at the heart of Dell's proposed solution to problems caused by increasingly dense servers. ... Intel plans to deliver in May its first dual-core chip for desktops.

... Rollins renewed his company's promise to work with Intel, despite a brief flirtation with Intel rival Advanced Micro Devices Inc. 'We saw Intel lagging a few months ago and expressed interest in AMD,' Rollins said. 'Since that time, Intel has really stepped up to the plate.' Now Rollins says it's not in Dell's interest to partner with AMD.' "

As you can see, Intel is following IBM's lead into multi-core designs. Injecting massive parallelism into the mix will be a little tricky since Intel has been trying to get that going with its Itanium project and that has not paid off well for the company – at least so far.

Interestingly, Dell's CEO announced that the company will use Intel instead of AMD because Intel has really come through recently. Yet, at the time of his speech, Intel's new chip was still not released and the AMD Opteron and Athlon processors continued to out perform equivalent Pentium 4 chips. So, you make the call. Was it technology or supply chain pressures that caused Dell to back off from AMD? Dell has never used an AMD offering in any of its PC models.

The facts about Intel's failure to deal well with the heat issue since 2002 are well documented. Meanwhile, the mushrooming demands for more processing power continue in the Intel chip space and Intel's inability to get rid of the heat to continue to exploit Moore's Law in single thread PC chips remains a big problem for the company. Dual core processors are not about to solve that any time soon. See Parallel Processing in Chapter 20 for the reasons.

How Intel Sees Its Future

To know where the future will take Intel, the best thing I can do is tell you where its top executive thinks the company is going in terms of technology and marketing. Nobody expects that a CEO at a major developers forum is about to say that the company was not as innovative as it once was and the future is bleak and the competition is better than it is – even if all of the above is true. When a CEO speaks, there is an implied "spin zone" and the executive is permitted to operate using euphemisms and positivisms – even if half the building is burning at the time. Intel CEO Craig Barrett at the March 2005 IDF in San Francisco was no exception to the CEO spin rule.

Despite its inability to conquer the heat problem in its Pentium line, after creating such a stir at the Spring IDF in 2002, in the three years that had passed, Intel's CEO had not gotten any less brash in his predictions. However, the Intel juggernaut has gotten smart enough to not suggest that its predictions are coming true any time soon. Intel learned quite a bit from its major 2002 IDF prediction that a 4GHz Pentium 4 was right around the corner. Almost three years later with major egg, Intel was forced to say that 4GHz Pentium 4's are practically unachievable and the company capped its line at 3.8GHz.

At the March 2005 IDF, Intel CEO Craig Barrett got out his crystal ball and just as in 2002; he again predicted that Moore's Law will boost chip abilities for many years to come. Barrett offered that the momentum would be kept up first through conventional manufacturing processes, then for many years after that by other technology. He then went on to predict that traditional chip-making technology would permit chip designs as small as 5 nanometers--about the width of 50 hydrogen atoms--to be used on processors.

Intel and its competitors in the industry today are preparing to introduce processors with features measuring 65 nanometers, or billionths of a meter. That means that 5 nanometers is way off for sure. Barrett said he was not sure what technology would be necessary to achieve the chips of the future but he was confident that Intel engineers would figure it out. "Meanwhile, it's full blast ahead with the standard Moore's Law," Barrett said.

Barrett went on to say that Intel chips are widely used in personal computers, but the company has failed in its bids to expand into the cell phone market. He noted that Intel has not given up on this market and he expected a number of design wins in 2005.

While he was listing some of Intel's problems in the form of opportunities, he chose to discuss another sore point for Intel – its Itanium processor family. As discussed in this book previously, this processor was once aimed at the entire market for powerful networked servers, but when that did not work, Intel subsequently earmarked the chip "for high-end systems."

Barrett kept Itanium alive at Intel and he said the company still has long-term plans for the chip. The CEO took the opportunity to push his company's support for WiMax, a protocol which can blanket a region as large as San Francisco with high-speed wireless networking abilities. "I think WiMax is going to be a disruptive technology that's going to change the way we think of mobile connectivity including mobile phones," Barrett said. "Hopefully, it will get us out of the half-assed broadband capability we have today in the U.S." Barrett is waiting for the protocol to be adopted by network service providers and he said he hopes that toward the end of 2005 or 2006 we'll all see massive commercial rollouts of this Intel capability.

Barrett chose not to recount Intel's problems with heat or its ongoing problems with providing higher clock speeds to its user base who expect higher clock speeds. He also did not touch much on being beaten by the AMD64 with Intel being forced to clone its own version called EM64T. In fact, he did not focus on the competition at all, like as if they were not really out there.

In an interview with reporters after his speech at the 2005 IDF, Barrett felt compelled to offer the upside of Intel's EM64T efforts. "We are out-shipping the competition in that space," Barrett said. It is worthy to note that this was a tacit acknowledgment that cloning a technology can result in increased sales. However, that's a role that CEO Barrett chose not to talk about

Intel began shipping chips with the dual 32 and 64-bit capability in April 2004 about a year after AMD's world leading, ground-breaking announcement. Barrett also noted that with 15 new dual core processors ready to be announced, he expected that 85% of Intel's shipments next year would be dual core.

In the overall market, Intel accounts for about 82 percent of all of the x86 chips shipped. AMD has most of the rest. Also, x86 chips account for more than 90 percent of the server chips shipped today. The rest are Sun and Power Architecture.

Regarding Intel per se, Barrett acknowledged the changing of the guard. In May 2005 he gave up the CEO title to Paul Otellini and he became chairman, replacing Andy Grove. So, for the future Mr. Ortellini has the CEO, President and COO roles to handle while Barrett assumed the chair.

As chairman, Barrett now participates in formulating strategies for Intel and will be the public face for the company when meeting with government officials, international dignitaries and educational leaders.

At the IDF, Barrett also noted that Intel will continue to garner more revenue and conduct more development overseas. "Seventy-five percent or so of our revenue comes from outside the U.S.," he said.

With a $40 billion plus revenue year expected in 2005, Intel's prospects for the future are bright. The fact is that the company is so dominant in the chip industry with its massive manufacturing capabilities and marketing clout that even if everybody decided to buy AMD in 2005, Intel would still get a huge percentage of the market because AMD does not have the fabs to build all the x86 family chips that the world needs. However, Intel should not relax for too long if it relaxes at all because AMD is part of a consortium in which the giant fab owners can very quickly begin to make AMD chips, in quantities more than enough to satisfy all demand.

That means that Intel can actually continue what some might call sloppy play as we have witnessed the last several years and the company will continue to be successful for at least awhile. Surely Intel does not expect to do that nor does the industry expect it to. The Intel Corporation is not interested in being embarrassed by AMD again and the company is certainly not interested in having the new plants that AMD is building or the capabilities of AMD's consortium partners take product shipments from Intel. One might suggest that having been whacked by heat problems and the superior Opteron AMD64 technology, this sleeping giant has been awakened.

Catching up in the AMD64 space and tying AMD in the dual core race tells me that Intel is not only not sleeping but the company is now quite vigilant and aggressive. There's only one thing that may hurt Intel. It has such huge resources that it can afford to make mistakes. AMD can afford no mistakes and it has made very few if any of note recently. Then again, there's IBM out there in yet another world, getting its guns ready.

Chapter 17
IBM in an Intel World

What is IBM Doing in Intel's World?

There is no question regarding the relevance of Intel in today's computer game. The company has had one major success after another and its successes are not about to end prematurely – even if it can no longer keep its promises. However, while Intel was floundering trying to jack up the clock speed of its Pentium and trying to get its second generation Itanium processor to appear worthwhile and easy to use, IBM was steadily moving further ahead in processor chip technology. IBM, a company that claims that it is not in a chip war with Intel or AMD quietly has become the undisputed leader in processor chip technology.

After having spoken to IBM in order to get my facts lined up for this book, it was very obvious that IBM does not want to be associated with the ongoing public processor wars of AMD and Intel. IBM does not want to be noticed in Intel's world and that is probably smart play. But, the facts speak for themselves. IBM is in Intel's world, though not in its face.

As a leader in producing powerful, low-power consuming, and cool chips, IBM has assisted AMD in achieving its own greatness in the industry with its 32-bit / 64-bit dual core Opteron and Athlon x86 processors. Perhaps AMD would have achieved its greatness without IBM's help but it is just too coincidental that shortly after its agreement with IBM, AMD began to use IBM developed technologies to best Intel in its own game. IBM knows how to design powerful, low-power consuming, and cool chips.

IBM is also a major contributor to Centaur's low-cost chip family? Centaur is wholly owned by Via Technologies and it too has major links to IBM. It's CEO, Glenn Henry, for example is an IBM Fellow, a mark of excellence in a scientist's career. Additionally, IBM is making Centaur's new 2GHz x86 chips at its East Fishkill NY fab and Centaur expects to be able to offer these chips at record shattering prices.

IBM is taking on Intel, quietly, under the radar, and now that IBM no longer needs to buy Intel chips for any of its products, Big Blue is acting much differently, though still quietly, and still apparently under Intel's wide radar net.

Every now and then, it helps to be reminded that IBM, from the 1950's onward, has lots of experience in trying to expand the limits of processing power with its own mainframe technology. Unquestionably, other than supercomputers, mainframes have been the computing power leaders for decades. IBM has always made the most powerful commercially reproducible computer chips for its own mainframe systems. Without having Intel as a supply chain constraint, IBM is now flexing its powerful muscles, but quietly.

Mainframes Deliver Power

Today's mainframe uses a CISC chip implementation rather than RISC and the mainframe chip is not generic as the POWER processors. It is specifically designed as a mainframe-only chip. IBM today calls its mainframe server line the eServer zSeries. With the small number of mainframe chips that are produced compared to all other processors inside of IBM, it is a very expensive proposition for IBM to have a separate chip methodology for mainframes vs. the rest of the company's servers, namely the eServer iSeries and the eServer pSeries. The latter models run on Power Architecture. IBM also makes servers that are based on x86 architecture chips from AMD and Intel and this line is called the eServer xSeries.

Though IBM tries to keep many of its plans secret, there is much industry speculation that when IBM announces its POWER6 generation of chips in 2006, as discussed in Chapter 15, the company will also adapt its mainframe line to use these chips. Since POWER5 came out less than two years ago, systems equipped with these chips have run twice as fast as the mainframe boxes. Once IBM can convert its mainframe operating systems to run on the POWER6 chips, all

IBM internal servers will be using POWER technology. It really says something about the advances in POWER technology that in the next year or two, IBM plans to begin to use the Power Architecture in its mainframe product line. The conclusion is obvious – there is lots of power in those chips.

IBM Leads the World in Parallelism

That's not all. IBM mainframes, pSeries, and iSeries servers support more users at a time than any servers that Intel has ever produced. While Intel is touting dual core processors, IBM has been making and using dual core processors since 2001 with the introduction of the POWER4 chip. While Intel offers a feature called Hyper-Threading on some of its chips, the POWER4 and POWER5 come with multithreading. Power5 in fact uses te more advanced notionof simultaneous multithreading (SMT). While Hyper-Threading has the potential to get two concurrent processes running on one chip, IBM's multithreading via SMT and its instruction pipelining techniques and its use of many devices on a chip enable as many as 200 instructions to be on the fly at the same time on one chip. Among other things, that's why IBM chips that run at 1.5 and 1.9 GHz consistently outperform Intel's 3.8 GHz chips in industry standard benchmarks.

In Chapter 15, we described the notion of IBM's mini supercomputer on a chip, the Cell processor. In Chapter 20 we explore parallelism in detail and in Chapter 25, we introduce the notion of Application Specific Integrated Circuits (ASIC). IBM is the recognized leader in customized ASIC chips that are found in every day items such as automobiles and Cell phones.

With its Cell processor and its POWER family, IBM is also the leader today in game consoles from Sony, Nintendo, and Microsoft. All of the three game consoles will use a derivative of IBM's Power Architecture and all will be built with explicit parallelism so that there can be multiple processor engines on the game consoles working in parallel to produce smooth motion video and exciting graphics.

Again, IBM has been doing all this quietly while Intel is just beginning to understand the magic of parallelism. Because of its technology lead, IBM is in Intel's world in the x86 area with AMD, Transmeta, and Via.

Moreover, there is nothing Intel would like better than to capture the game market and the ASICs market for Cell phones. But, Intel is getting beat by IBM's technology in every area that the chip giant's marketing clout cannot control its customer's actions. When given a real choice, Intel's customers and potential customers seem to have no problem leaving Intel for the best technology – quietly.

IBM is the Leading Processor Technology Provider in Intel's World

IBM is the industry leader in parallelism, RISC, ASICS, multi-core, multithread, and 64-bit computing and it is all working under the covers of the Power Architecture chips. Additionally, the IBM Cell processor tops the charts in all computational areas. On top of the company's internal efforts, IBM has taken much of this leading technology to AMD, and in this respect, Big Blue along with AMD lead the Chip Giant even in its own x86 platform.

Chapter 18 The Heat Is On

The Exception to Moore's Law

It is well documented in Chapter 16 that Intel has been suffering over the last three years (2002 – 2005). If not for its inability to achieve major processor advances using the implied performance corollary of Moore's Law, then because it had boasted that it would do so. On top of its failures to overcome the heat problem in its Pentium 4, Intel's Itanium problems and getting beat by AMD64 have all contributed very negatively to its development and marketing efforts in the processor chip area. The only really positive press for Intel was that the company was able to convince Apple, despite all of Intel's recent failings to hop on board the Intel chip line.

Meanwhile, IBM's POWER processor chips have been going gangbusters and IBM has even had the time to help Intel's # 1 , #2, and #3 enemies, AMD, Centaur, and Transmeta with some substantial advances in their microprocessor lines. Is it coincidental that shortly after IBM's agreement with AMD, the company was able to introduce its powerful 64-bit Opteron chip with the 32-bit, x86 compatibility feature? IBM had been doing 64-bit chips for eight years before the Opteron and six years before Intel brought out its Itanium.

AMD, not IBM competes directly for Intel's chip business. In an open market, AMD with IBM's help may eventually be more than Intel can handle. Some might suggest that Intel is merely reaping what it has sown. After all, the company predicted that Moore's Law (corollary) would rule but a few hot flashes along the way kept the Moore predictions from materializing this time. Unlike IBM and its POWER technology chips, which focus on massive parallelism of up to 200 instruction streams on the fly, Intel has relied almost solely on Moore's Law for increases in the clock speed of its Pentium class chips. But the law broke down at the heat sink.

As noted in Chapter 16, at the Intel Developer's Forum (IDF) in late February 2002, the Intel Company boldly proclaimed to its loyal constituency that Moore's Law was alive and well and that Intel would continue to apply the law to gain speed in its processors. In essence Intel said that it would continue to reduce the space taken by circuits on chips and would effectively double the number of circuits per silicon chip every eighteen months, and – and this is the big one, the implication was that the speed increase would be attainable with no sweat – just like all the prior bumps. This corollary no longer holds true as explained in detail in Chapter 21.

The essence of Moore's Law, and its performance corollary, named after one of Intel's founders, Gordon Moore had been paying huge dividends to Intel and nobody in Intel seemed to believe that the corollary to the Law, which Gordon Moore never acknowledged as being related, was about to expire. You may recall that at the forum, Intel went on to predict that 4.0 GHz versions of its Pentium 4 chips were just around the corner even though its fastest chip at the time was running at clock speeds of about 3.0 GHz. Surely, great things were expected to continue to happen according to the corollary.

The Heart of the Problem

Now, let's get to the heart of Intel's problem. For many years, Intel's chip architecture had relied on halving the circuit space, and tweaking the cycle time to increase processor clock speed. Doubling the clock speed again and again from its first chips to the present using Moore's Law had always delivered as promised. With Moore's Law continuing to pay off in terms of circuit densities, Intel can certainly package twice the amount of circuits per chip for years to come. However, as the company learned the hard way, Intel does not have the technology to cool the chips so that they can actually run at the higher clock speeds. Thus, higher speed Pentium 4 processors and their follow-ons are usable no place other than locations close to the Sub-Arctic. Actually, there are big problems as you can see in Chapter 21.

More than Just a Heat Wave

Heat has become the issue with today's processors especially Intel's Pentium 4 processors. When Intel released its Pentium 4 Prescott version of the chip, many in the industry recognized it for what it was-- a heat machine.

In 2004, Intel was wrestling with heat and the company was working with heat and cooling vendors trying to get the hot Intel chips cooler.

How could heat become such an issue out of the blue? It was like a shock to Intel. In fact, in 2004 at the IDF Steve Pawlowski, Director of Intel's Microprocessor Lab admitted that at Intel it was always about performance, performance, and performance. Intel had always focused on producing the fastest chips possible in the given time. The heat never worried Intel before "because it was not a concern."

Where Is the Big Clock Speed Boost?

Intel actually has a bigger problem than what is on the surface. The computer enthusiast community likes to take its boxes and alter the clocking to get even more speed from the processor. With Intel's Prescott at the cusp of failure due to heat, there is little room to ramp up the speed. In my personal unit with which I am typing this book, periodically I get a heat warning from Windows XP telling me the processor is hotter than the threshold. So, if I am getting heat warnings at 3.2 GHz, what about the 3.8 GHz machines and what about the folks who drive x86 CPU speed up for their own personal use on their own machines. It is safe to say they should expect failure.

By driving up clock speeds, without adding other technology, boxes run hotter and hotter. Since Intel has always been driving up clock speeds to achieve power boosts, one would think that the engineers would have observed the heat phenomenon and that they would have planned for it. Why doesn't AMD have the heat problem? Why doesn't IBM have the heat problem? Perhaps the answer is because IBM recognized the problem long ago and has always designed around its potential for failure.

Heat Choices?

When Intel became acutely aware of the heat problem, the company began to use stop-gap techniques to solve it. The Chip Giant began to use improved case fans and heat sinks to dissipate heat in a more efficient way. There are not really many inexpensive choices, however. Finally, Intel took the logical choice. The company began to limit the clock speed / the frequency of the chips to reduce the amount of heat that is produced. Ironically, limiting the clock speed and frequency will not do it for Intel anymore. Users want more power.

Only after three years of not being able to solve the problem did Intel finally admit to having problems with heat. Some analysts think that this is the company's first right step to solving the heat issues for the future.

Nobody questions that the engineers at Intel are brilliant. They have consistently raised the bar on performance using a law and an implied corollary that appeared would never be broken. Yet, this esteemed group received a physics lesson in early 2002 that lasted to 2005 as they tried to stretch the speed of Pentium processing further than the heat generated from firing up all the circuits would permit.

Many scientists are now suggesting that heat has become the most important problem in CPU chip design. Chips can be much faster if the engineers can find some effective ways to reduce the waste heat. Unable to do this, engineers must be content with more efficient systems to conduct the heat away from the CPU chip and perhaps do something as straightforward as release it to the room. If only it were that simple. One thing is for sure, Intel has yet to figure it out.

Until 2002, the Intel approach had worked very well and had kept Intel chips at the top of the performance game. Now, in 2005, the IBM approach of multi-threading on the chip, multiple cores, and parallel processing has gotten the highest form of flattery from Intel – imitation. Intel has announced that its future chips would be multi-core and they would focus on getting more things done at one time rather than merely relying on higher clock speeds.

Intel is starting to learn – heat!

Chapter 19
IBM Knows Heat

Blazing Mainframes

Those of us who have been in the business for some time can remember when IBM's behemoth mainframes were so blazing fast that its fastest computer systems required real plumbing to provide water cooled heat dissipation. This made building the systems expensive and it made installing the systems a challenge since both plumbers and electricians had major roles in preparing the compute room facilities.

I bring in the mainframe as an example again to strengthen my argument that at this time, IBM engineers and scientists understand the heat issues with processor chips better than those in any other company. Big Blue has been solving the heat problem in chips for decades as its mainframe processors have always been pushing the power and heat envelopes. That's why the IBM Company holds so many patents for its research and it has many solutions to the nagging, ongoing heat problem in building fast processors. Heat is not a new problem.

Back in the late 1980's IBM solved some of its heat problems with the introduction of the water cooled thermal conduction module shown in Figure 17-1. IBM later created air cooled thermal conduction modules and the company has been recognized as a leader in tackling the heat problem in chips and modules from the first time that heat was detected as a major enemy of high speed computing.

The water cooled thermal conduction module as shown in the picture below was a phenomenally revolutionary and equally clever innovation in heat dissipation technology. By reading the snippet under the picture, taken from IBM's Web site, you can get an appreciation for the processor power, heat, and packaging problem that IBM had solved using this most unusual design.

Figure 19-1 IBM Water Cooled Thermal Conduction Module - 1987

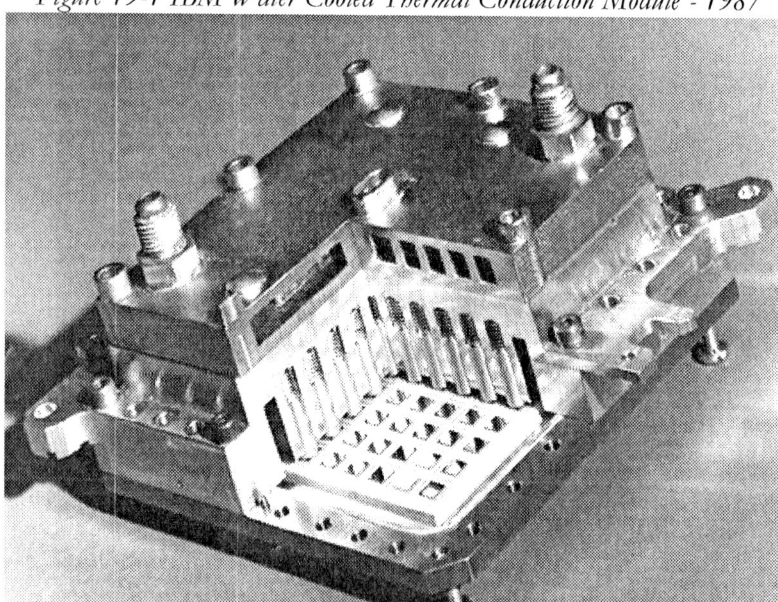

"By 1987 the information processing power of electronic computers of the 1960s could be held in one hand. The Thermal Conduction Module seen here was the marrow of the large IBM 308X computers. Six inches square, the TCM had room for up to 133 chips, each with 704 circuits. Each ceramic block of the material had 28 to 33 differently wired layers. More than 350,000 holes provided paths for the vertical wiring for layer-to-layer communication. The chips were joined to the substrate through a total of nearly 16,000 contact points, using IBM's unique chip-joining technology. All that circuitry generated 300 watts of heat -- enough to destroy the chips. But the heat was drawn off through spring-loaded aluminum pistons (seen in the cutaway section) that pressed gently against each chip. In turn, the pistons were housed in a "hat" filled with helium, an excellent heat conductor. Chilled water flowing through a conduit attached to the hat whisked the heat away. One TCM alone -- there were about two dozen in a 3081 computer -- packed as much computing punch as a medium-size System/370 of only a decade before."

Cooling with Air

In 1991 IBM took its TCM technology one notch further. With its new Enterprise System/9000 type 9121 Model 320 Processor (the IBM mainframe of the day), IBM introduced an air-cooled thermal conduction module (TCM). The fabrication of this module required the integration of a litany of new chip types including new bipolar chips, CMOS SRAM chips, and ECL and DCS logic circuitry. All of this was packaged in a TCM that could dissipate heat by means of air cooling.

The water cooled TCM of 1987 and the air cooled TCM of 1991 demonstrate IBM's leadership in heat technologies dating back almost twenty years. At the same time, in 1987 Intel was still running cool with its 386 chips and in 1991; Intel continued to be cool with its first iteration of Pentium chips. Intel had no need to face the problem.

Heat Dissipation and Other Techniques

Because IBM has been studying and dealing with heat and how to best cool hot chips for decades, it is safe to say that IBM knows heat. In fact, IBM is not only a leader in heat dissipation technology but over the years, the company has been developing techniques on the chip itself through embedded software to help the chip run as fast as possible without overheating. One of the major techniques that IBM has used successfully for a number of years in its processor chips is parallelism at the chip level with multiple cores, efficient pipelining, multithreading, and smart shutoff.

Moreover, because parallelism in particular can be difficult to implement outside of an engineering lab, IBM has developed on-chip techniques to efficiently handle the difficult task of creating threads dynamically at execution time, rather than forcing programmers at the source level to change software to define threads and run code in parallel. The benefits and the problems of parallelism are further explained in Chapter 20

IBM Heat Work

As noted regarding its mainframe work, IBM has been working with heat as a major problem for many years. In the last five years, IBM has instituted major innovations in all of its chips to help cool them down so that less heat is generated.

Low Power Products Initiative

The following snippet gives a perspective on a new heat dissipation initiative in IBM as it was kicked off in 2001.

IBM Launches "High-Powered" Initiative for "Low-Power" Products

http://domino.research.ibm.com/comm/pr.nsf/pages/news.20011001_lowpower.html

EPA Recognizes IBM Mainframe as Industry's First Energy-Efficient Server

"Armonk, N.Y., October 1, 2001 . . . IBM has launched a company-wide initiative to improve the energy efficiency of information technology for enterprises and consumers, establishing a worldwide low-power computing research effort to be coordinated out of its research lab in Austin, Texas.

The company also has established a low-power consulting practice and is accelerating the development of ultra-low power components and power-efficient servers, storage systems, personal computers and ThinkPad notebook computers, among others.

'Very quickly, energy and heat will go from being irritants to major product development limitations,' said Mark Dean, Ph.D., IBM Fellow and vice president of systems research. 'The demand for increasingly powerful systems is driving up the amount of heat within many new products. If we don't address the power issue, products will become so hot that you'll be able to cook with them rather than compute with them.' "

Dean has been named to coordinate the low-power initiative on a worldwide basis for IBM. He is serving as the focal point for ongoing efforts at the Austin center and other IBM research facilities as well as with IBM product groups and customers to speed high-performance, power-efficient products to market.

Cool Blue

When you have lots of R&D, you never know where inventions are going to pop up. On July 12, 2005, IBM announced a new product called "Cool Blue." This technology component can be used with the existing chilled water supply for air conditioning systems already located in the majority of customer datacenters to reduce server heat emissions by up to 55 percent.

Cool Blue is installed as a door that hinges to the back of a rack. It has a hose installed in the floor that goes up the door. Sealed tubes inside the "door" filled with chilled water remove up to 55% of the heat generated in a fully populated rack. Cool Blue then dissipates the heat by pulling the hot water into the unit for cooling so the heat is not released into the data center.

Where dowes all this heat come from? First of all the chips run hotter as we have been discussing. Secondly, a good part of the heat problem in data centers is caused by the proliferation of blade servers – made by IBM itself as well as other vendors such as Dell and HP. This has created major heat problems. The small size of a blade server in a rack permits data center managers to pack multiple servers into a rack space formerly reserved for a single box, making the idea of a special cooling unit to augment air conditioning a practical addition.

IBM is always researching heat and ways to avoid it and ways to get rid of it, so it is not unexpected that IBM would be the first systems vendor to develop such a technology. "Cool Blue" is universal as it can be deployed on any server, enabling customers to ease the burden on existing air conditioning units and even lower energy costs by up to 15 percent.

This unique design is for servers not chips, but with the very hot servers that are being produced today, the demand for heat dissipation technology in data centers has jumped dramatically. This innovative uinit is designed to easily fit on to the IBM eServer Enterprise rack or just about any rack. Managers whose datacenters have reached the limits of cooling capacity, but still have space to add racks of systems are the ideal candidates.

The improved cooling from the Rear Door Heat eXchanger enables data center managers to fully populate individual racks, freeing valuable floor space without the need to purchase additional air conditioning units. The Heat eXchanger can also alleviate the issues caused by other equipment in a datacenter where there are "cooling challenges."

IBM has been addressing the cooling needs of customer data centers for years and the new Cool Blue' technology is another leap forward in overcoming previously insurmountable air conditioning limitations. As customers try to incorporate more processing power into the same data center footprint, this breakthrough technology will help them win the war on heat.

Of course, IBM's first battle is to reduce the heat on the chip. Even though this invention is not at the chip level it again demonstrates that IBM not only "knows heat," but IBM is doing something about it in all areas all the time.

IBM Seees Moore's Law Limitations

In its POWER chip line long ago, IBM recognized that Moore's Law has its practical limits where heat is concerned. IBM's designs recognized that chip power would be limited by how well heat dissipation technology was able to keep up with the heat generated by faster and faster processors. Realizing that the heat generated from processors ramped up to high speeds would cause the chips to become unreliable; IBM chose a different approach than Intel.

Whereas speed increases for Intel had depended mostly on increasing the clock speed on its single thread chips, IBM's chip designs concentrated on performing more functions (more work) in parallel at lower clock speeds. Thus, IBM's techniques produce less heat. For example, according to Dr. Soltis, when using the POWER5 processor, as many as 200 instructions can be in flight in one chip at one time. IBM's philosophy of massive parallelism, rather than single threading at high rates of speed has paid off in better heat management. In 2005, IBM's POWER5 chip was clearly the most powerful chip in the industry in terms of being able to get work done. Because IBM knows heat, the POWER5 chip gets its job done and it is cool.

Chapter 20

The Problems with Parallel Processing

The PC Performance Gravy Train is Gone

For those who absolutely depend on a faster PC every few years, the gravy train has pulled out of the station and there is no new train coming. The software industry has become dependent on a 35-year drive to increase clock speed and, that drive has run out of track. This hurts Windows developers especially, because their single-threaded applications (execute one instruction after another in series), will not run any faster on a multicore chip than on a single core Pentium or Xeon. The industry is in for a shock, and people are starting to realize that a lot of Intel applications will need to be re-invented in order to run any faster at all on new x86 PCs or servers.

Since there are no more Intel clock speed performance increases on the horizon, the way the new chips are supposed to give a performance boost is through a notion called parallelism. Some level of parallelism is designed into all the new chips from AMD and Intel so it is not a notion that will escape the PC user the next time they need a newer and faster PC. But, because most programs today are what are called single thread, without changing software, new PCs are not expected to boost performance very much, and in some cases, they may actually slow down your old reliable software.

What is Parallel Processing?

The idea of parallel processing, or parallelism, is very simple to define. It may not be that easy to use, however, depending on the design of the processor chip involved. Simply put, in computer terminology, parallelism is the ability to have more than one program or parts of a program use the resources of one or more processors (cores) on a chip at the same exact instant.

A simple analogy would be a carriage drawn by one horse compared to two or four. All four horses would be operating in parallel at the same instant with each handling ¼ of the entire load. The load of course in this analogy could be four times greater than it would be with just one horse in harness.

Intel has two architectures today that support parallelism. One is its newest line, the Itanium series and the other is the Pentium line also known as x86. Itanium is completely new and uses a new form of processing than Intel's standard x86 architecture as found in its Pentiums, Xeon's, and Celerons.

AMD's Athlon, Opteron, and Sempron models also use the familiar x86 style of computing and thus suffer from many of the same limitations when running in parallel as Intel's chips. However, because AMD has been working with IBM to incorporate various IBM-developed parallel processing and other performance techniques into its chip designs, AMD dual core chips as a rule get more work done than Intel's. AMD leads the x86 industry today in processor power in many ways because of its efforts with IBM. IBM is the recognized leader in parallel technology. IBM's dual core POWER5 chip, for example, which runs at less than 2GHz is substantially faster than Intel's 3.8GHz offerings and the chip inherently supports a "no-sweat' parallel processing implementation

There are five main problems with Intel's implementation of parallelism, multi-threading, and multi-core processors, and there is yet another issue if you are entertaining the Itanium processor:

1. Existing single thread programs get no benefit (Most Windows programs are single thread and Windows has been a single thread operating system).

2. For improved performance, Intel's parallel technology requires programmer work or new software.
3. Because they haven't had to do it before, Intel programmers don't possess the necessary skills.
4. Programmers want to spend their time adding function, not performance.
5. If Intel changes the chip architecture, programs will need to be rewritten AGAIN.
6. Itanium is an animal by itself.

1. Existing Programs Get No Benefit

What about all the programs that have specifically been written for Windows or Linux over the years with no concern for parallelism?

A quick check would demonstrate that almost all Windows and x86 Linux programs today are single thread. What does that mean to somebody buying a new server or a new PC? Well, if you need one particular program to run in less time, you may be just as well off not changing hardware. With nothing more than Hyper-Threading or dual core processors running at the same 64-bit GHz level as the non-threaded, single core unit that you are replacing, a single application will more than likely run no better.

There will be little payback on the dual core and Hyper-Threading chips for many users. Therefore, industry analysts are predicting major disappointment by computer purchasers upgrading from relatively new technology.

Single Core to Dual Core

It helps to recall that parallel technology had no real play in the Intel world until 2004 when AMD forced Intel's hand to match its AMD64 architecture and its pending dual core offerings. Most existing Intel code, and there are mountains of it, in both Linux and Windows environments has been written to depend on the 18 month shots of speed delivered by clock frequency improvements based on the performance corollary Moore's Law. To achieve this periodic increase in speed, developers and users never had to be concerned about directing code to be run in multiple processors or writing code for multiple threads within one processor. They simply would purchase the machine with the higher clock speed and the Intel chip would deliver the performance boost first time, every time. That paradigm is finished and already Intel is feeling some aftershock.

As described in Chapter 18, the high clock speeds that developers and users have been enjoying for decades are now associated with intolerable amounts of heat and the need for top notch heat dissipation technology. Compared with Intel, instead of big clock speed boosts, IBM's efforts with its chip technology since the early 1990s have focused on getting more things done in one tick of the clock-- as many as 200 in – flight instructions can execute in parallel in IBM's Power Architecture. When a chip gets lots of work done in a single cycle, the cycle speed is not as much of a determinant as the work that gets done during the processing cycle.

The difficulties involved in Intel moving its Pentium 4 and Xeon chips to 4GHz have compelled all chipmakers to reassess their devotion to raw speed and turn attention to other ways to improve performance. AMD quickly found that the 64-bit processor would help quite a bit, doubling the data bandwidth and offering far more addressable memory space. Now, the x86 industry has embraced the notion of dual-core processing that (combined with dual threading) should bring additional performance gains to multithreaded applications. But, Intel does not accomplish parallelism as IBM does in the Power Architecture. Therefore, it delivers no benefit to single thread applications without lots of work.

How Can Programs Run Better?

How can you or I as Intel users be able to take advantage of the new parallel hardware technologies with minimal pain? You may be able to purchase your programs again from the software vendor, but only if your vendor has rewritten the programs for parallelism. If the programs have not been re-written, then your programs just won't run any better than your old model at similar clock speeds and cache levels. This is not going to make you and I and a lot of other users very happy.

Those who have trusted Intel over the years to provide single thread performance boosts with clock speed increases are in for a big frustration. Without clock speed increases your single thread programs (just about all programs you have) will not take advantage of the new parallel hardware that Intel is placing on its processor chips.

The market reality is that massive software rewrites are almost impossible. So I predict that eventually the users who are not tuned into the drawbacks of single thread software running on systems that can support parallel processing, will be let down but they will eventually accept the reality. Initially, users will expect a job that ran on a 3.8 GHz Pentium to run better with two 3.2 GHz processors. However, in most cases, the job will actually run slower.

One might expect that the next great advance for x86 chips is for AMD to take additional advantage of its IBM partnership and begin to add the on-chip capabilities to its own multithread and dual core chips. If AMD x86 chips begin to be designed like more like IBM POWER chips and parallelism is on-chip rather than developer driven, AMD will emerge as the only viable alternative in the x86 space to a software re-write. Of course Intel can take the same route also but, for a company at the top of its marketing game that likes to go it alone that is highly unlikely.

2. Intel's Parallel Technology Requires Programmer Work

Depending on whether the computer chip is designed to assist in setting up parallel streams in hardware while the machine is running or whether the chip requires pre-built parallel software, there is a major difference for a software developer. As many things about computers, there is ample computereeze to categorize these two variants. If the parallelism is perfumed in hardware on the chip, for example, we would say that the chip has "dynamic hardware thread optimization." If the work for parallel processing is done instead by the developer using a software tool known as a compiler, then, we would say that the chip uses compiler level thread optimization.

For a chip that uses compiler level thread optimization, a software developer (programmer) would have to endure a high degree of difficulty to provide the machine with an optimized program with which to work. Compared with the typical "single-thread" programs with which most programmers are accustomed, the difference is as dramatic as the difference between night and day. The task can be quite daunting. It all depends on how the chip was designed.

To be technically correct, a chip that performs its threads dynamically in parallel activities typically also gets some help from a compiler. In other words, even though the chip itself may provide for instruction level parallelism and/or thread level parallelism as well as a number of other dynamic tools for parallelism; most computer makers, including IBM, also provide some level of assistance in their compilers. The key difference here is that the programmer is not involved in the creation of the threads, or in any manual optimization scheme.

To repeat, the degree of difficulty for the developer lies in whether the chip maker used hardware or compiler thread optimization. Though this first sounds like gobble-d-gook, it is actually easier to understand than it may first sound. Let's break down the three operative words: (1) Hardware in this notion means that the work is done on the chip with native chip instructions rather than being done in software such as Windows. (2) A compiler is a piece of software that translates instructions written at the source by a programmer into machine language that a computer can understand. (3) A thread can be thought of as an executable entity of work in the system. It may also be viewed as a separate instruction stream.

Intel Multithreading

To enable parallelism, a chip is equipped with a number of transistors to handle parallel processing. These include multi-threading hardware and/or multi-core hardware A Pentium D chip for example has two processor cores and it has something that Intel calls Hyper-Threading, which provides the ability to run two threads in parallel per core. Theoretically, in a perfect world, a Pentium D or a dual core Itanium (late 2005) should be able to perform four times as much work as a single thread, single core Pentium 4 or Itanium, respectively.

The combination of the dual thread hardware and the dual core processor makes it appear to the operating system as if there are four distinct computers under its control. Whether or not four times the work gets done is a function of how well the processor chip is optimized – either dynamically in flight or ahead of time by the compiler. As you might expect, the objective in parallel processing is to optimize the hardware components on the chip that are built to run in parallel. Because of other performance related factors, however, the measured effect of Hyper-Threading with Intel chips is not 400%. In fact, it is more in the neighborhood of a 20% to 30% improvement. Moreover, the user gets few to none of those benefits unless the software is reworked to be translated by the multithread compiler.

IBM Multithreading

IBM has been doing multithreading on its POWER chips since 1998. In 2001, Big Blue upped the ante when it announced the POWER4 chip with two processor cores and a full set of registers to support multithreading. In 2004, POWER5 refined the POWER4 approach with a design known as simultaneous multithreading (SMT). The IBM design for POWER5 parallelism was just about as good as it gets. Whereas, the Intel technique of Hyper-Threading typically achieves a 20 to 30% improvement over single threading, and only after a programmer fixes the code, the POWER5 design allows each core on the chip to behave like two processors running at full speed.

IBM took three additional years to churn four times the power from POWER5 than POWER4 without an increase in circuitry. POWER5 gets its SMT abilities through a different use of the POWER4-designed execution units. These are the parts of the chip responsible for executing the various types of instructions. With the improved design, each POWER5 chip behaves as if it has four separate processors. Depending on the application workload, POWER5 is therefore able to increase performance up to four times that of POWER4.

3. Shortage of Experienced Intel Parallel Processing Programmers

Developers in need of parallel capabilities seem to love Intel's compilers. Since Intel knows its chips better than any other source, the company builds a thread-capable C++ Language Compiler and a

FORTRAN Compiler and offers tools to its developers for them to be able to set up these programs in multiple threads. Intel uses a compiler-based method for EM64T (Intel's x86 technology) and Itanium.

Compiler Level Multithreading – The Intel Way

Using this approach, the developer is responsible for the arduous task of identifying threads for the compiler and then telling the compiler how the threads should interact. All of this work must occur before the processor ever gets to run the program as a single threaded unit. This is a tricky proposition and it requires substantial work for the resultant code to gain an advantage when running. Comparing the Intel approach to the dynamic approach in IBM's POWER line, we might suggest that IBM's approach is like a 15-speed bicycle bought already assembled versus one with "some assembly required."

The Intel compilers may be an acceptable solution for the few software developers with heavy threading experience. However, there are not many of these, especially in the Intel camp. As we discussed previously in this book, Intel has just emerged from being the clock speed champ in which most programs have been written for a uniprocessor, are single threaded, and have depended on clock speed increases for performance boosts. Therefore, a typical Intel software developer has little to no experience in parallelism and has not been trained how to work with threads at a compiler level.

For Intel this is not good news. The programming talent for parallel processing for Itanium or for x86 has not been cultivated and just is not there today in the Intel world.

4. Programmers like Productive Work

Another point of reality is that few program developers want to have to work in an area not directly related to coding business function into software. In other words, they are not interesting in figuring out how to multi-thread using compiler tools.

Coding for threads is extra work. Additionally, even if there were a talent pool to write new programs, and also develop the threads for new applications, what about the exiting investments in x86 software?

There are not many experienced Intel parallel programming experts because for years programmers required no knowledge of parallel processing techniques for their code to run acceptably on Intel hardware. With the new dual core models, however, that is no longer the case. I am not suggesting that the old code won't run on the new 64-32-bit dual core processors, it will run fine in 32-bit mode but it won't take advantage of the other threads or cores that may be available on the chip. In other words, on a new machine, it will not run better.

Over time, programmers will learn to adapt their code unless Intel or AMD offer a better alternative and that is possible, but they may do so kicking and screaming. The fact is that many programmers will not want to optimize their code for the chip once they have gotten the job done. If the program works, many programmers feel their work is complete. Lots of programmers will not be interested in splitting a fully functioning program into threads that can run in parallel even if they are forced to learn how to do it.

Dynamic Speculative Multithreading– The IBM Way

Wouldn't it be nice if you were using a processor chip that did not force you to rewrite or buy replacement software just to use the benefits of the new chip? In the Power Architecture, IBM has decided that programmers do not want to keep rewriting their code to move to better performing chips. So, IBM does the work for you. Along with instruction level parallelism and pipelining, IBM uses a feature called *speculative multithreading* and out of order execution and it creates threads and/or instruction streams dynamically (on the fly) at the chip level. In this way, the programmer does not have to work as hard and the programmer does not have to rewrite software in order to gain the power of the next generation POWER chip.

Most Power Architecture chips are used in a multiuser environment in which as many as several thousand users can be running the same or different programs from just one one chip. In these cases, with the help of the operating system and native chip virtualization schemes, POWER processors can have many functions going on in parallel at once, without the programmer or the user being involved in the set up.

Fill Up the Instruction Pipeline

Before we continue the discussion of IBM v Intel parallelism, I want to briefly explain another notion of parallelism, namely, instruction pipelining. The explanation is not fully technically accurate but it is simple so that you do not have to understand all the parts of the system in order to understand the important notion of instruction pipelining. See the Hint below:

> **Hint:** Instruction Pipelining Another common component in modern microprocessors is the instruction pipeline. This technique provides a form of instruction parallelism in that the instruction stages that must be completed while executing several instructions may be done in parallel. On its way to execution, an instruction goes through various stages. Let's say for our simple processor, there are four distinct stages: The instruction (1) must be fetched, (2) loaded in a machine register, (3) decoded and finally (4) executed A pipelined processor may be able to work on four different instructions at the same time, with each instruction at a different stage. On modern RISC processors, all of these tasks can be taking place simultaneously in the "instruction pipeline" so that four or more instructions can be in various stages of execution at the same instant.

Instruction pipelining and multithreading are both supported in IBM's Power Architecture. Multithreading is a popular topic nowadays because Intel has announced what it calls Hyper-Threading, and the company has been making it available the last few years in its top end Xeon processors and more reently on Pentium. Compared to IBM's use of SMT and pipelining, the net effect of Hyper-Threading on system performance has paid off for Intel users in the 20% to 30% range. The reason this is not a higher percentage in practice is related to the work that must be done outside of what the operating system provides. The fact is that lots of Windows code gets no advantage from Hyper-Threading.

Programmers Do Not Always Guess Correctly

As you would expect, in order to have lots of things going on in a parallel processor, (1) either a human being in the form of a programmer with a compiler has to break the programs down into smaller tasks, streams, or let's say; threads or (2) the chip performs the function dynamically on the fly. In building software for Hyper-Threading, and/or dual core units, it is conceivable that a programmer, without complete vision of how the program will behave during execution, may make an erroneous presumption about what can run well together. The compiler may also presume erroneously. In some cases, it is likely that things may actually run slower in a poorly conceived parallel environment and / or they may even bomb in execution.

A programmer's guess while using the compiler may not be correct. The programmer cannot anticipate what will actually happen in a dynamic multiprocessing environment but with compiler optimization, when the program is running, it is too late for the programmer to help. If the chip does the optimization work at execution time, however, and it makes a poor decision, it has current execution state information and thus, it can then backtrack down the pipeline and re-optimize the current environment dynamically. The chip never has to ask the programmer for advice because "it" knows more about the execution environment than the programmer. To run this thought home, I am compelled to offer the following modified line from Erich Segal's 1971 movie, Love Story: "[Hardware optimization at the chip level] means never having to say you're sorry."

To permit programmers to work a regular work-day, IBM has built special hardware onto its POWER4 chip to make all this good parallel stuff happen without the programmer caring what's going on. Programmers, thank you, need not be involved. IBM's special techniques are at the leading edge of parallel technology. A quick list of IBM's POWER4 developments include the following:

- ✓ Speculative superscalar outoforder execution design
- ✓ Up to eight instructions can be issued each cycle
- ✓ Sustained completion rate of five instructions per cycle
- ✓ Register rename pools for virtualization
- ✓ Other outoforder resources

- ✓ 15 stage pipeline structure
- ✓ Over 200 instructions in flight at any given time
- ✓ Eight full execution units
- ✓ Two identical floating-point execution units
- ✓ Maximum 4 floating-point operations (FLOPs)/ cycle / core
- ✓ Two loadstore units,
- ✓ Etc.

POWER5 is even better. Please do not think you need to know the items on the above list. I present these here because the list of facilities is impressive, and there are more. Using these on-chip tools, IBM has enabled as many as 200 instructions to be in process in POWER5 in one core of one chip in one instance in time. IBM's chips pre-fetch instructions and try to analyze the branches ahead of time and create instruction streams / threads on the fly. With each new chip, IBM's algorithms get even better. Right now in most cases, 90% of the speculative branches are actually taken. If you contrast the dual threading technique implemented by Intel, you can easily see how much more elegant and capable, the IBM technique is. Plus, and this is the big one, it requires no additional programming. Then again, IBM has been doing this for years and Intel is new to the parallel game.

To help those who have a penchant for understanding the dirty details, I am providing a word for word explanation of how these things work together directly from the IBM POWER Design Guide:

> Hint: The internal microarchitecture of the core is a speculative superscalar outoforder execution design. Up to eight instructions can be issued each cycle, with a sustained completion rate of five instructions. Register rename pools and other outoforder resources coupled with the pipeline structure allow the design to have over 200 instructions in flight at any given time. In order to exploit instruction level parallelism there are eight execution units, each capable of being issued an instruction each cycle. Two identical floating-point execution units, each capable of starting a fused multiply and add each cycle, i.e., a maximum 4 floating-point operations (FLOPs) per cycle per core, are provided. In order to feed the dual floating-point units, two loadstore units, each capable of performing address generation arithmetic, are provided.

5. Changes to the Chip Architecture Causes Rework

As a general rule, it does not make good business sense for application programmers to be forced to write programs requiring a deep knowledge of the hardware architecture. Intel has recently released two x86 compilers – one for C++ and the other for FORTRAN. So far, that's it and that's all there may be. If you write Visual BASIC, Visual RPG, Visual COBOL, "Visual" Java or other languages, even if the language supports threads, you may not get any performance improvement if you take the time (substantial) to make your program thread-aware. The company that builds your compiler has to know the chip as well as Intel knows the chip.

But, let's say that you do choose to work at the thread level with your compiler and you take the time to split things up and you spend twice the time of what it would take to write single thread code. Then, you implement and voila, it works. Depending on the compiler you use, you may get a 25% boost in performance from Hyper-Threading or you may not. The program may not work any faster depending on the reality of the compiler and the chip at the time. Worse than that, after all that work, depending on your knowledge level and hardware skills, it may not even work as well. On top of that, a future version of the compiler or a future chip may take away any benefit you have achieved and to get it back, you may have to rework your application again.

WordPerfect as an Example

When DOS WordPerfect was written to run on the 8088 in the early 1980's, for example, it was a wonderful program. Then, the WordPerfect Company wanted a GUI DOS version and they rewrote the program so it looked nicer. They had tied the original code into a specific physical interface and when they changed it to home-spun GUI, the code had to change. In other words, they had to rewrite lots of WordPerfect to make the change.

Then, of course Windows 3.x had its own GUI. To use the Windows GUI, the WordPerfect developers had to rewrite the code to run under Windows and they had to rip out the GUI part and do it all again. When Windows 95 came out, it was so different, that for the fourth time, the

WordPerfect developers had to gut their code and make it run well in this new environment. The 32-bit Windows 95 application was updated to run on Windows NT and Windows 2000 and Windows XP but this was not a big effort because the code stayed at a 32-bit level through all those iterations. There were no major hardware changes.

Now, while all these versions of WordPerfect were being released, the WordPerfect Company was paying huge dollars to get the new version out as quickly a possible. They were not about to provide their software in its new version form for free. So, through each of these iterations, the personal and corporate user community had to cough up the cash to buy a new version and had to learn the nuances of the new version

Now, here we are with 64-bit computing and with multithreading on the horizon. This means one or two more rewrite and / or re-buy options for software. To go to 64-bit computing, you must rewrite your code and to go to Hyper-Threading (HT), you must rewrite your code. So, who knows about the next Intel architectural change that must be done to gain performance benefits? Each time that the developers of say, Word or WordPerfect have to rewrite their code, one thing is for sure, you get to buy it again. If they choose never to come out with a 64-bit version or a real multithread version of the software, then the old 32-bit version will run, but it will not run better.

Are all systems like this? No, they are not! The IBM POWER5 chip, for example is in its fifth generation. It runs on AS/400 machines, iSeries machines as well as IBM's new i5. In Chapter 23, we discuss a notion called holistic design in which the whole system is considered in the design of the chip and all components. At a software level, the predecessor of the i5, announced in 1978 as the System/38 was designed in a holistic fashion from the ground up.

There is not enough room in this book for me to explain the architecture of this advanced system, but let me say this. Ninety-Nine percent plus of the software written for the System/38, a CISC 48-bit hardware unit in its day (1978) is portable to the i5 without a programmer having to touch the code. Programs do not even have to be recompiled. When the object code arrives on the i5, it immediately is re-encapsulated to run on 64-bit RISC hardware (POWER5) and the application programmer is not involved in the process. The old software immediately functions with multiple users using multiple threads-- automatically in the new environment.

6. Itanium Is an Animal by Itself.

Intel is not really new to parallelism since, even before threats from AMD, along with HP, the company has had its own parallel processing effort underway with its Itanium and EPIC (Explicit Parallelism) projects. These projects have been alive for over ten years but have not paid off in any measurable way for Intel or HP. In fact, HP begged out of the Itanium project in 2004 as the prospects for a marketing success waned. It has not been an easy time for Intel either. Other than this one major project, and dual threading, Intel had not tried to introduce parallelism into its x86 Pentium line until 2005.

With its "Hyper-Threading technology" that had been deployed on the company's more expensive Xeon chips and lately on its Pentium 4, two threads can be alive at the same time in a single processor core. In 2005, Intel introduced its dual core chips providing up to 4 threads of parallelism on its x86 line. In April and again in June 2005 the chip giant demonstrated its Itanium dual core models due out by the end of 2005, and the company announced that quad core Itaniums were scheduled for 2007.

In the initial batch of mainstream Pentium "D" models in 2005, however, the Hyper-Threading feature was conspicuously absent as Intel was still getting the kinks out. Intel rushed out its new models to get the dual-core technology on the market quickly so that AMD would not gain ground. Hyper-Threading was available only on the more expensive Intel supercharged dual-core models right from the start. More than likely Intel will roll it into the Pentium D and other Intel dual core chips over the next year or so.

Other than these initiatives, Intel, unless we include Itanium, which many still consider experimental, has not had much of a leading seat at the parallel processing table. Since the bulk of Intel's real parallel work had been with Itanium, quite frankly, that really means that most of its work so far has been unusable.

Unlike IBM's efforts with parallel processing and RISC processing, in which the chip creates the threads dynamically during execution, Intel and HP made a critical design mistake at the outset of their Itanium partnership. The partnership decided that the application developer (programmer) would have to be involved in determining the various threads and how they would be compiled and executed.

Because the reverse engineered EM64T technology did not include chip-level thread optimization, and there was a rush to get the EM64T based chips out the door, as noted above, Intel is now in the unenviable position of asking the world to change its software to be able to take advantage of the parallel features of its newest chip offerings.

The IBM PowerPC approach is much simpler for developers as it takes regular code written in a regular way, with no required knowledge of threads, and at the chip level it is optimized for multiple processes. The programmer does nothing differently and that means that IBM's parallel technology is as usable for performance gains as Intel's single thread paradigm of continually ramping up clock speeds.

The following snippet from IT Week discusses the Itanium alternative to single thread Pentium processing. (Multi-thread x86 processing has the same flavor.) It does a nice job of putting the dilemma of Intel and HP (now just Intel) regarding software conversion to Itanium [and x86 multithread) in its proper perspective. In fact, it may be one of the reasons why HP abandoned the Itanium project.

http://www.itweek.co.uk/analysis/1141320

Comment: Intel faces an Epic struggle

"Dell no longer sells Itanium systems, IBM has put the chip on hold, and even HP seems to be hedging its bets on Itanium and its 64bit Epic architecture, says Roger Howorth"
Roger Howorth, http://itweek.co.uk

IT Week 02 Jun 2003

"Intel makes much of the fact that Itanium's Epic architecture enables the chip to efficiently run suitably optimised software. This is because the Epic design does not use hardware on the chip to optimise software as it is running. Instead, Itanium assumes the software was optimised when it was made. This is different from other chips, which

> use hardware to dynamically optimise software while it is being executed.
>
> But the problem with compiler-based optimisation is that software runs best when it is optimised differently for each version of the chip's architecture. There have been two versions of Itanium architecture since its launch; no doubt more are planned. Intel argues that upgrading to a new version of the architecture does not mean users will have to update the software because a new chip will run at a faster clock speed.
>
> … The Epic approach seems to signal server software upgrades every year or two.

Though publicly Intel has not given up on Itanium, its lack of software compatibility has sure given the chip a bad rap. After a disappointing start, Dell stopped selling Itanium boxes. Microsoft stopped working on the workstation version of Windows for Itanium. The worse blow, however is that HP, which inveested tons of money and co-developed Itanium with Intel pulled out of the Itanium business completely in 2004. After over 10 years of development, Intel and HP seem to have made a costly chip that is suited to very few niches.

To make a long story short, the Itanium and its notion of parallel processing so far is a marketing bust. HP is out of the deal and many analysts believe that Intel will more than likely fold the whole project. The fact is that because Itanium requires a software rewrite before you even get to the trauma of coding for Itanium's unique brand of parallelism (VLIW), it is not a desirable x86 replacement. Forcing programmers to rewrite code to use Itanium and then to optimize their own code to achieve the benefits of parallelism is way too much to ask programmers who had become accustomed to buying clock speed as needed.

As Intel changes from single thread processes to the IBM type of performance maximizing methodologies that conserve heat, software developers are going to have a hard time keeping up with the changes required to gain the performance benefits available on the chip. Unlike the automatic benefit gained by faster clock speed, Intel developers now are faced with lots of work in order to gain from the new Intel parallelism effort. It's not free. I predict that Intel software developers will look back kindly upon the good old days when the 3 GHz Pentium was at the top of its game and software just ran faster as the chip speeds increased. "But that was yesterday, and yesterday's gone."

It's not Tomorrow, It's Now!

The move to dual-core processing has occurred faster from a chipmaker's perspective and more easily than many IT professionals might have expected. Windows is actually quite adept now at supporting Hyper-Threading and multiple processors. Though there are not many multithreading versions of many applications available, those that are show great promise.

IBM has had dual-core chips for almost five years, and has had on-chip multithreading since 1998. Additionally, IBM's new POWER6 and even more powerful processors are on their way. In mid 2005, both Intel and AMD, for the first time, brought their mainstream x86 dual-core processors to market. More and more dual-core products will appear next year and quickly move to the forefront of IT. Itanium is set to join the dual core Intel line by year end, 2005.

Summary

Regardless of whether the Chip is a dual core AMD Opteron, a dual core Intel Pentium D model, or single core Itanium, to use the multi-threading facilities, the programmer will have to be involved.

To summarize, let us reexamine in summary form the six main points that we have discussed in this chapter.

1. Existing single thread programs get no benefit (Most Windows programs are single thread and Windows has been a single thread operating system).
2. For improved performance, Intel's parallel technology requires programmer work or new software.
3. Because they haven't had to do it before, Intel programmers don't possess the necessary skills.

4. Programmers want to spend their time adding function, not performance.
5. If Intel changes the chip architecture, programs will need to be rewritten AGAIN.
6. Itanium is an animal by itself.

Existing Programs Get No Benefit

There are billions of dollars invested in x86 software for Windows environments and for Linux. In a compiler optimized multi-threading environment, old programs will run no better and may perhaps even run worse since only one of the cores will be used to run the job and more than likely the clock speed on the chip (both cores) will be less than the single thread uniprocessor that had run the work previously.

Intel's Parallel Technology Requires Programmer Work

Unfortunately, to achieve performance boosts with the new dual core and HT Intel Chips and AMD chips, you must plan for parallelism. You can either rewrite your programs or buy new ones

Shortage of Experienced Intel Parallel Processing Programmers

Since experienced Intel programmers (Windows, Linux) have never needed to invest in other ways to make their single threaded applications run with multiple threads, they are not now equipped to get the job done. Since Intel always came through with a clock increase in the nick of time that required no programming work, programmers never needed the additional skills needed for parallelism.

Itanium is a double whammy for Intel. (1) Because the chip uses a new technique for parallel computing called VLIW (Very Long Instruction Word) computing, there is no similarity between Itanium and the traditional x86 programming model so programmers must convert code or write new for Itanium. (2) Since the Itanium method of parallelism is optimization by the developer at the compiler level, and not many Intel people have experience with parallelism, optimized Itanium applications will be hard to come by.

Programmers Like Productive Work

Writing code for threads is much different than writing code to solve a business problem. The fact is, once a typical x86 programmer solves a business problem or writes a snazzy game for the x86, the last thing they want to do is to have to rework the job so it runs OK on the system. So, not only are x86 programmers not trained for the job, they don't want it.

A lot of money may get spent on dual core systems that provide little to no benefit.

Changes to Chip Architecture Cause Rework

Even before the introduction of dual core processors and parallelism, Intel EM64T and AMD's AMD64 customers had a problem to face. Because AMD provided backwards compatibility on the AMD Opteron and so did Intel, 32-bit programs (almost all programs are 32-bit) continued to run. Because they made some architectural improvements to use the 64-bits in the operating systems, user programs that were 32-bits ran somewhat better with no additional work on similar GHZ machines.

However, if these workstations or servers actually need the performance boost of a 64-bit machine, the programs must be reworked to run in the 64-bit environment. This is in addition to the changes to make the programs work in the parallel environment. Moreover, if Intel or AMD switch to 128-bits in the distant future, this change will be forced on the user community again in order to take advantage of the new architecture. With IBM's POWER technology, such costs are avoided.

Itanium Is Way Different

Whatever the workload it may be to change code to run on the 64-bit and/or dual core x86 processors from AMD or from Intel, it may be a veritable picnic compared to writing code for Itanium. The Intanium processor is a totally different architecture so any x86 program that knows anything about the hardware must be rewritten. Since Itanium is a parallel processor by design, the next step after getting software to run is to get it to run well using the tricks of compiler parallelization. Good Luck.

Conclusions

Clearly dual-core is the latest trend in processor development. However, it is no panacea. It is conceivable, for example on the x86 dual core entrees that some applications may actually run slower than on a single-core processor. If for example, you run on a single processor 3.8GHZ Pentium 4 and you move to a 3.2GHz dual core model, without optimization, the program will run slower even though there is another processor sitting idly by. Applications with many threads may experience performance problems when faced with slow thread availability. Then, of course there is the age old problem of poor I/O bandwidth. If there is not enough bandwidth available to support the type and quantity of devices to which the threads are talking, the system will slow down.

In order to best utilize dual-core Intel processors, users must have modern operating systems running such as Windows XP and it would help to look for applications that are threaded such that they can take best advantage of the hardware enhanced threading capability provided in Intel HT processors, dual-core based Intel processors, or AMD's flavors.

Will the Second CPU Sit Idly By?

If additional power is delivered through a second processor as opposed to a faster first processor, then it may not be the developers who complain, it may be the users who need more power than is available in one core processor to run their job. In dual core processors for example, two jobs can run simultaneously but neither job runs any better than it did when it ran on a single processor.

So, developers will have to do something differently in their code in order to gain the speed improvements using the AMD64, EM64T or the Itanium way. That means that software will run but not in the second processor unless the developer writes it to run in multiple threads. As discussed previously in this chapter, in its POWER platform, IBM uses smart multithreading on the chip. The chip itself looks ahead and creates threads and instruction streams on its own without developer anticipation or intervention. That really means without the developer having to do any additional programming work.

With Itanium, and multithreaded x86 systems, the developer must anticipate potential threads while coding the program and then split the application manually at the source level. First of all, this is a tedious task and secondly, if the developer anticipates the threads improperly, the application may actually run worse, not better.

With its agreement with IBM, we should expect that Big Blue will teach AMD how to make its chip work the way IBM's works. So, unless Intel adopts the notion of on-chip smart multi-threading as opposed to source determined multi-threading as in its Itanium, its developers and users will more than likely be disappointed when they begin to use the "next generation" Intel chips. The second CPU will be sitting idly by.

Power Architecture Is Amazing

IBM's POWER5 processor chip is not just dual-core; it is equipped with an enhanced version of Power4's SMT (Simultaneous Multi-Threading) facility. Though similar to Intel's HTT (Hyper-Threading Technology), IBM's POWER SMT is far better in that it dynamically optimizes parallelization by analyzing and prioritizing threads to make parallel execution much more efficient. For example with just a .35 GHz increase between POWER4 and POWER5, and with the same number of circuits, the POWER5 runs four times faster.

POWER5 adds two basic thread-prioritization schemes to the SMT mix - dynamic resource balancing and adjustable thread priority. So, the POWER5 processor analyzes the behavior of code as it is executing and it prioritizes some code and it sidelines other code that could slow down the SMT stream. For example, instructions that must be executed in sequence to derive an accurate result can lock that thread in the processor for a time. POWER5 tries to predict this and run simpler instructions until there's room to execute the sequence without clogging the SMT hardware.

The application-defined thread priority can be issued by an operating system or hypervisor in concert with the chip and is a major factor in dynamic resource balancing calculations. This function is necessary to determine the length of time a thread remains active in the CPU. It also gives operating systems an easy way to control power conservation.

Running a slew of high-priority threads will cause the chip to run hot. But as the hypervisor or the operating system knocks thread priorities down, the CPU will run more idle cycles and therefore run cooler. When all the thread priorities are knocked down to their lowest level, the CPU goes into a sleeplike low-power mode. One might consider that the simplest approach to power management that can be imagined.

POWER5 also uses a smarter from of power management as it can power down portions of the chip that aren't needed at that moment. This potentially puts a new spin on the PowerPC's "infamous power and heat problems" that Apple so willingly put the spin on last year. One would think that with all this power management facility built in to POWER and not x86, Apple may have a hard time in the future using its operating system driven power management schemes (if they exist) as employed by x86 processors.

Final Thoughts

This chapter may very well have been tough reading for the neophyte but there are some basic principles of parallelism that knowingly or unknowingly will affect all PC users in the next year or so. If it is any consolation, let me assure the reader that it was no tougher than it was on me. I rewrote this chapter three times to make it more understandable. I can assure you each time it got easier to read. I hope you enjoyed it.

I'd now like to close this chapter with a quote from Tom Yeager writing for InfoWorld in a December 2004 article titled: "IBM's

POWER5 worth a second look." Tom says it as good as it can be said about IBM's secret weapon – its arsenal of POWER chips:

"POWER5 has got just about everything: speed, simplicity, innovation, seamless backward compatibility, a mature development toolset, and the backing of a technological giant. It's an unrivaled engineering achievement, created by what may be the world's smartest engineers. If IBM's marketing ever matches the intelligence of its engineering, watch out, Intel."

Chapter 21
Is Moore's Law Dead?

Observations in Moore's Law

Webopedia defines Moore's Law as:

> "the observation made in 1965 by Gordon Moore, co-founder of Intel that the number of transistors per square inch on integrated circuits had doubled every year since the integrated circuit was invented. Moore predicted that this trend would continue for the foreseeable future. In subsequent years, the pace slowed down a bit, but data density has doubled approximately every 18 months, and this is the current definition of Moore's Law, which Moore himself has blessed. Most experts, including Moore himself, expect Moore's Law to hold for at least another two decades."

To get a better feel for exactly how this is to be interpreted, let's look at a few definitions for the word "law," and the word "rule." These are taken from Dictionary.Com. There are many others.

Law:

1. A rule of conduct or procedure established by custom, agreement, or authority.

2. A generalization based on consistent experience or results: the law of supply and demand.

Rule:

1. An authoritative, prescribed direction for conduct

Since the definition of Moore's Law is an observation, then we can conclude that it is not a rule of conduct or procedure established by custom, agreement, or authority since the circuit density on chips never agreed to any rules and they do not behave by custom but by invention. Thus, Moore's Law is a generalization based on consistent experience or results as much as the law of demand and the law of supply. Of course, the objective in this analogy is not to teach economics but to contrast the notion of an observation of something that merely happens without causality to something that is caused to happen by a rule.

Quite simply, the law of demand states that, if all other factors remain equal, the higher the price, the less people will demand a good. In other words, the higher the price, the lower the quantity demanded. This is not a rule, it is an observation. There will be people who choose to buy at a higher price – just not that many in aggregate.

The law of supply is the inverse. It states that the higher the price, the higher the quantity supplied. Producers supply more at a higher price because selling a higher quantity at a higher price offers greater revenues. Again, this is just a generalization based on observation. It is not a rule since any one supplier can choose to sell at a reduced price.

To complete the economics picture before we make more sense of Moore's Law (observation), the law of supply and demand indicates a relationship between the two in that when the supply function and demand function intersect at a certain price, the economy is said to be in equilibrium. At this point, the allocation of goods is theoretically at its most efficient because the amount of goods being supplied is exactly the same as the amount of goods being demanded. At the given price, suppliers are selling all the goods that they have produced and consumers are getting all the goods that they are demanding. This too is not a rule, it is a conclusion based on observations.

So, Moore's Law is an observation. It was a very astute observation and until densities drifted to the 18 to 24-month time period, the observation was 100% accurate. When density increases began to slow, the rule was redone to fit the new observations. Moore's law did not cause any density improvements - ever, but it was and is a good predictor of densities. Good engineering and the masterful use of basic and advanced physics is what helped the industry achieve its phenomenal chip densities.

Vince Said It Would Be Like This

We have a weatherman here in Wilkes-Barre, Pennsylvania whose name is Vince Sweeney and he works for WBRE TV. He is quite a performer and a good "weatherman." The station would love to have the general public believe that there is a cause and effect relationship with what Vince says and what actually happens weather-wise. It's a cute little marketing gimmick. When Vince hits it on the nose, the station loves to interview the locals and have them smile and say to the camera: "Vince said it would be like this." Of course, Vince did not cause the weather to be nice but he did "predict" the weather from past observations and other factors just as Gordon Moore originally predicted circuit densities doubling every year.

Nothing is perfect. So, whenever a cloud brings a pile of rain on a sunny day, Vince updates his prediction. Likewise when densities began to double in eighteen month to 24-month time frames, Gordon Moore updated his observation (prediction,) which for marketing purposes at Intel, today is referred to as a "Law."

Intel's Vested Interest in Promoting Moore's Law

Just as WBRE has a stake in the locals believing there is a cause and effect relationship between what Vince says and the actual weather – since viewers may be more prone to watch the channel if they accept that connection, Intel has a big stake in the scientific community and the public believing that Moore's Law is causing all these circuit density breakthroughs. Moore's Law has been and continues to be a good observation of the only variable Gordon Moore ever tied it to – densities. Any other observation after that is purely speculative.

Now, what additional observations am I referring to? From the first time an increase in processor speed was observed with the introduction of a denser processor, Intel and many others attributed the processor speed increase to Moore's Law. So, for over thirty years, Intel itself would ramp up the clock speed each time a density breakthrough occurred and the chip would do more work. The chip would run "faster." Moore never made a law tying density achievements to performance but Intel and the rest of the industry had no problem attributing the performance to Moore's Law. In this book, we have referred to this as the performance corollary to Moore's Law.

While Moore's Law continued to be correct in observing density breakthroughs, something bad happened to the implied performance corollary at about the 2002 time frame. Intel could not ramp up its clock speed to 4GHz as promised because the performance corollary failed. Thus, from that point forward, nobody could say that improvements in density were directly proportional to achievable increases in chip performance. Moore's Law, however, was not broken as the densities were achieved. In fact, Moore's law still is alive and well. But, without the Intel performance corollary, it is just not exciting any more. In fact, as IBM's Meyson says (Chapter 22), it is irrelevant.

The engineers at Intel or its marketing people or the CEO knew or should have known that adjusting just the length and width density factor on a chip (Moore's Law) and ramping up clock speed was destined to fail as a long-term strategy. Density is not the only factor for performance and Intel learned a hard lesson in 2002, from which it has yet to fully recover.

So, now that we no longer have Moore's Law as the guiding force to assure power, what is the next step? How can faster PCs and x86 servers come to market without faster clock speeds. The answer is one word, parallelism, but as we learned in the last chapter that too has its issues.

Chapter 22

The Winds of War: IBM and Intel

Pot Shots over the Bow:

While AMD is negotiating with the department of defense to strengthen its offensive capabilities in its war against Intel, IBM has been quite silent. Well, most of IBM anyway. At the 2004 Design Automation Conference (DAC), the premiere conference for engineers in the semiconductor industry, Intel CTO Pat Gelsinger could not help firing a pot shot at IBM's CTO Bernie Meyerson for apparently and allegedly not pledging allegiance to Moore's Law.

The 2004 DAC Conference

Though IBM has told me many times in my discussions with company personnel that "the IBM Company does not compete against Intel," I have concluded that IBM's official posture is merely a ruse. It is a ruse for the public and for Wall Street and for Intel. I'm not really sure who is buying it, but I certainly am not. Wall Street isn't doing much with it if it is aware. Intel as a company seems to be playing along but from his remarks at the 2004 DAC Conference, Intel's CTO Pat Gelsinger isn't really buying it either.

After some amenities and a look at his impressive background, Gelsinger shifted his keynote address at the 2004 Design Automation Conference (DAC) in San Diego to a big plug for Moore's Law and a big "shame on you" if you are not a believer. I could almost hear him say, "Vince said it would be like this." (See Chapter 21)

Attendees of this conference are the engineers and the bigwigs working in the semiconductor industry. I got the feeling that Mr. Gelsinger was being just a bit disingenuous when he made this emotional appeal to this audience. By insisting that Moore's Law is not dead, and it isn't, Gelsinger was out there pretty close to begging the argument with reasons other than two dimensional densities, as to why Moore's Law was so important to his audience

"Moore's Law is the engine that drives our industry."

Most analysts would suggest that technological improvements through innovation drive the semiconductor industry, and being able to miniaturize circuits on a chip is certainly a big part of that. An observation however cannot drive anything. So, the beginning of Mr. Gelsinger's message to DAC was to defend Intel's assertion that one of its founders, Gordon Moore created a law that drives the semiconductor industry. If you are Intel, it is good to posture that an Intel Law drives the Industry. But, if you are any other semiconductor company, you know that Moore's Law gives a great estimate of how far technology can go in achieving densities. However, Moore's Law does not cause those densities, nor does it drive the industry.

Gordon Moore observed a phenomenon that through classical scaling, certain densities were being achieved and he saw no reason why those high densities would not continue to be achieved. With or without this observation, however, the densities would be achieved. That's the point.

Following this claim, Gelsinger plugged Moore's Law a bit more and then he continued to associate Moore's law with scaling and suggested that there are those who believe that scaling (according to Moore's Law) is dead. As he continued this emotional trek, there was something in Pat Gelsinger's craw and he had to get it out. So, he went after IBM's CTO Bernie Meyerson by name and basically branded him as a Moore's Law skeptic, a chronic non-believer. Then after he settled down, he gave a very impressive speech.

This was the first volley in the public fight between Intel and IBM. The subject matter has to do as much with pride as it does with anything else. The whole world of Intel users had taken Intel's lead and had associated performance gains with Moore's Law, a notion which, as explained in Chapter 21, is untrue. When the performance gains stopped cold even though the regular improvements in densities continued, Moore's Law was not about to cause Intel processors to run any faster. Intel had cranked up the clock one time too many and the company got caught with a fire and lots of egg on its face.

If Moore's Law as Mr. Gelsinger highlighted is the same as classical scaling then once tweaking the density and cranking the clock speed failed to work, Moore's law became an irrelevant notion in performance. IBM's Meyerson counters Pat Gelsinger by flatly stating that "Moore's Law is irrelevant." Interestingly enough, after having Pat Gelsinger in 2004, the folks at DAC invited Dr. Meyerson to be the keynote speaker at the June 2005 conference and yes, there was a returned salvo.

Intel's Gelsinger Keynote at DAC

Before we present Dr. Meyrson's position on Moore's Law, I have transcribed the beginning of Mr. Gelsinger's keynote for your review since it was the first public shot across the bow. From Intel's side, at least for pride and intellectual bragging rights, the war of words with IBM is on. Eventually, this war will turn to a public marketing war but my guess is that the marketing war itself will be going on for a few years before either side admits it.

> "...Many have suggested that the era of our industry is over. That the period of Moore's Law Scaling as we've described it has come to an end. And when you look at an exponential graph such as this, when you look at lithographic costs, and you look at CAD costs, when you look at Test costs, all of those things are increasing at a very astounding rate
>
> And against that, one of the questions that I get asked more than anything else as Intel CTO is: Is Moore's Law Dead.?
>
> Have we come to the end of this period -- because Moore's Law is the engine that drives our industry? If we stop having a larger canvas to do our designs on -- Hey CAD tools are just a commodity just like everything else that we do in our industry the pendant engine continues to look forward. Hey these are what we call the unfriendly exponentials of our industry.

> And in fact some have said that scaling is dead right --- and these aren't arbitrary people; these are people who know what they are talking about. I'd like to argue very vehemently that you know in contrast to Bernie Meyerson, of IBM's CTO --- or anybody else who argues against this, I fundamentally disagree – Moore's Law is alive and very well.
>
> As a technologist, I disagree. As an Intel person, I disagree. But as a businessman, I am actually delighted that other people think this is falling away because that gives us [Intel] a bigger opportunity looking forward
>
> Fundamentally, believe in the law."

Mr. Gelsinger went on to deliver a brilliant speech about the issues in the semiconductor industry and was warmly received by the audience. However, the shot across the bow was noticed by IBM in the person of Bernie Meyerson.

The 2005 DAC Conference

As expected at the 2005 DAC Conference, Dr. Bernie Meyerson, who delivered the keynote, had his day. However, unlike Pat Gelsinger at the 2004 conference, Meyerson mentioned no names. But everybody knew who he meant.

By the way if you want to hear these presentations in their entirety, go out to www.dac.com and click on the presentation that you want to hear, Dr. Meyerson or Mr. Gelsinger. They are quite interesting and in many ways their salvos have fueled the great intellectual battle between IBM and Intel. The title of Meyerson's talk went right at the heart of Pat Gelsinger's challenge from 2004:

How does one define "Technology" [chip technology] now that classical scaling is dead? (and has been for years)?

It did not take Meyerson long to get at the matter at hand but to help warm up his audience, he did take a stab at humor by recalling the picture of him that Pat Gelsinger had implanted in his PowerPoint Presentation the year before when he was chiding the unfaithful, such as Meyerson about Moore's Law. Here's how Meyerson shot back – one year later:

> "Normally you start one of these things by thanking people for the invitation but I've got to express some concern about your judgment. From what I understand, last year you saw a ten foot high photo of me -- with a caption that read
>
> "Bernie has love child with Martian"
>
> – and you invite me anyway – an act of bravery"

Dr. Meyerson had obviously been preparing for his keynote for some time and he chose to meet head on all the issues regarding Moore's Law, but most notably perception. In the abstract advertising the speech for DAC attendees, he told them exactly what to expect:

> Over the past four decades, the IT industry has relied upon the classical scaling of semiconductor technology to drive both performance and product economics. Often confused with Moore's Law, classical scaling speaks to the science driving performance gains over the past decades, not the subset economic issue addressing the real density of transistors on a chip. In effect, classical scaling had been the "glue" binding microprocessor economics, as stated by Moore's Law, to expectations for ongoing progress in microprocessor performance. The impact of the loss of that linkage with the demise of classical scaling has yet to be fully comprehended. The discontinuity engendered by the failure of classical scaling has shaken the microprocessor and IT industry to its foundation, forcing radical shifts in product roadmaps and business focus for those unprepared. This talk will briefly review the origins of this discontinuity, but more critically emphasize new strategies, such as Holistic Design, as employed to drive continued progress in IT performance. First results of the movement to Holistic Design, at chip and system level, will be reviewed, as well [as] strategies meant to accelerate efforts in this vein.

For the most part, his talk was about scaling the Intel way and scaling the IBM way. In this next section from his speech, you will see some more humor but then quickly back to basics as Meyerson went right back to Moore's original canonical expression to provide insights:

> "Now, the title [title of slide] is meant to get your intention but it is a very important fact. We are living through what is arguably one of the greatest transitions this industry has ever undergone. There are a couple of questions here.
>
> You know this is the Peter Paul and Mary question in terms of quotes from Hitchhiker Guide to the Galaxy – it's roughly the same thing --- Where have all the GHz gone? That should get the attention of the design community.

In his less than subtle style, Bernie Meyrson played with the notion that the clock speed and the density ratios have pulled apart and regardless of densities, the clock speed boosts have ended. This is not a shot at Moore's Law as we explained in Chapter 21, but it is a muted shot at Intel, the company that knowingly permitted itself to benefit from the confusion between density and performance. I know that Bernie's weapons were not loaded with real bullets, however, because IBM does not permit its people to say what they really think in public – when they are representing IBM. But, this was a shot nonetheless.

> "We spent the last 40 years looking at frequency as the driver in this industry and all of a sudden you never hear people talking about it anymore. That should have your undivided attention and you have to ask yourself have the skills diminished or is there something more fundamental at work?
>
> "Now, I want to go back and start at the beginning.
>
> "Ya know Gordon Moore was one of the great geniuses of this field and published this paper starting by the way with the word "cramming" – Cramming more components into integrated circuits.
>
> "In 1965 what he said was with unit costs falling as the number of components per circuit rises, by 1975 economics may dictate squeezing 65,000 components on a single chip of about 1 cm sq or less. If you do the math -- what really it means is this – you double the density of what you can put on a chip in a fixed dimension roughly speaking every 12 to 18 months.
>
> "This is an economic law; it is a law about the density of a chip. It is not a law about the performance
>
> "However because of the success of Moore's Law and the accuracy with which it predicted the future, we fell into a pattern where basically

we tended to think small and smaller when it comes to microprocessors and the reason is simple:

"Small transistors are faster and cheaper. We assume they use less power. They allow more function on chip. They are phenomenally predictable. I mean think about it. You could predict in the past ten years forward how fast a transistor would go with scale and power and it was pretty accurate. They enabled progress and drove this business for forty years. However, you have to remember how they were created. They were created using something called 'classical scaling.'

"To understand the problem, what I want you to do is this, take a piece of paper 8 x 11 and fold it in half ten times Before you go to bed tonight - Those of you who are larger, meaner, and faster than I am – please don't do it!

"You'll find somewhere around seven folds things get exciting. My point is that if you do something over and over and over again it stops working.

"It's kind of like for all these years we had what I call the small knob in this industry. But if you turn a knob enough times (4 decades is probably enough!), it breaks. The small knob broke.

"What do you do? The small knob broke.

"First of all we have to define a few terms

"I have always said that classical scaling is dead – not Moore's Law

"Classical scaling is the 'synchronous reduction of the entire set of device parameters, not just the area of the device' [defined by Moore's Law] It is a hugely complex issue-- far far far beyond the complexity associated with Moore's law.

...

"The fact of the matter is there are about 20 parameters [or knobs] that you must pay attention to in scaling, unless this is a 20 dimensional universe. Take my word on this it is more than just x an y – [as defined by Moore's Law.]

"So this has very little to do with Moore's Law. In fact, all this says is: How do you scale a transistor without causing it to self immolate, which is a rather important thing since if it is your laptop, you're not going to be happy with that...."

Bernie Meyerson went on but he had already made his point. Moore's Law is not dead. Circuit densities will continue to improve and probably at a Moore's Law pace. The supposed continual performance improvement link that had always been associated with Moore's Law is gone in its original context, for sure. For more than three years and counting, Intel, the industry's largest chip maker was not been able to scale using what Meyerson calls "classical scaling."

Where Have All the Gigahertz Gone?

In one of his slides, making his point with more humor, Meyeson asked, "Where have all the Gigahertz gone? That's a fair question. Not only have I not seen any gigahertz lately, Vince stopped saying "it would be like this." The implied correlation of Moore's Law and users going "Wa Hoo -- Is that fast!" is dead and was buried in x86 land in 2002. Yet, many have still not noticed and that is Meyerson's point.

So, why did it all break down so fast? Where have all the GHz gone? Let me offer this simple thought. When you are well, you feel OK. When you are not well, in most cases, you do not feel OK. When you actually get sick, there is a period in which you still feel OK though your blood might not pass all the tests that it normally would. You might say to yourself that "just a minute ago I felt fine." Or "I've been getting worse all day." The fact is that the moment right before it is no longer OK, it is OK. So, the Gigahertz left town when classical scaling failed to produce a chip design that could perform under the same circumstances as before. Forty years later, at Intel, the heat brought down the implied performance corollary to Moore's Law.

Balanced System Performance

As a point of contrast, IBM has been fine tuning all areas of its systems since it started to build systems years ago. Chips are just a part of the system. For example, the IBM System/3 was a real dog in terms of its mathematical and scientific abilities. However, it was built to be a business machine. So, IBM chose not to waste dollars on a fast floating point processor for a machine that was not going to be doing floating point.

On the other hand, eventually IBM attached an 1100 line per minute printer to the system. This was needed to produce business reports rapidly. IBM provided a nice wide data path for this printer and it built the system with something called print spooling so that when the printer was ready to print, it would not have to wait for something slower on the system to complete.

Without getting too much deeper into this, it would have done IBM no good to run a System/3 Model 15D with a blistering fast Pentium 4 processor. If retrofitted to 1969, the processor would have been waiting for the printer and would have been well underutilized. This is not balanced performance. IBM called its System/3 a balanced performance system – all the parts were balanced to work with each other. So, a slow processor was all it needed to get the job done. The BlueGene/L supercomputer is another example. It is the world's fastest supercomputer, yet each individual processor runs at just 800 MHz, or .8GHz. That's balanced performance.

Bernie Meyerson could have told the folks at DAC that IBM's POWER processors are balanced chips. There is no sense ramping up clock speed if the wires are too thin or thick or too resistant. There is no sense ramping up the clock speed if the gates will overheat and create current leakage and power loss. Unless all 20 or so design items on Meyerson's list are optimized (not maximized) for the overall chip environment (IBM now like to call this the ecosystem.) in which they will live, then they are prone to failure. Though it took forty years for classical scaling to create performance and heat issues and leakage issues, the day before there was a heat issue, before the ramp up in clock speed, there was no heat issue. That's; why nobody, including the leaders at Intel even cared. They did not see it coming. The only other explanation is that somebody said to ignore it.

IBM, Intel, AMD or whoever have to look at the whole chip in order to know how best to optimize all its pieces. Moore's Law still predicts density breakthroughs accurately however, it has no bearing on whether the transistors and interconnects and the gates on the chip will all function well at the new reduced size. And, when the die size hit about 130nm Meyerson asserts that the gigahertz were hard to find through classical scaling. That is what happened to all the GHz. They are gone.

To make an impact on the crowd at DAC, Meyerson used a striking analogy. He started by asking again "Why is scaling dead?" Then he answered his own question as he said, "Because Atoms don't scale." He said hey, let's say gate width or a critical distance is say 5 atoms thick and we reduce the technology to double the density. Does this distance become 2.5 atoms? Do we start a nuclear fission?

He gave the analogy of placing your hand in a 55 gallon drum of sand and pressing down. There should be a nice handprint there. Sand, in this context, scales well. Everything moves nicely to displace the sand where the imprint is. Now, he said, fill the barrel up with bowling balls instead, and repeat the test. Not so nice. All of a sudden, it doesn't scale well.

Meyerson also made a major point of the importance of keeping the power density constant in new chips so that more heat is not created. He quipped: "Why is constant power density critical?" and then as expected, he answered his own question:

"Because a 1 million watt laptop doesn't work."

He noted that there is no real plan B to make up for the lack of GHz -at least at Intel and he offered that the only thing left is innovation. As you look at how IBM is positioned for the future, you must conclude that Big Blue is uniquely prepared to innovate to overcome all the problems of scaling, performance, and chip cooling. Meyerson says that in this new age, you have to schedule invention to meet future roadmap objectives. IBM has seen the scaling problem coming for years and has had its research teams working on many different plans B. Just because IBM does not give away all its secrets does not mean there is not a bucket load of secrets in the vault with more being developed every day for the future

IBM is ready for the challenge and one of the techniques that the company is using to be successful is Meyerson's notion of holistic design.

Figure 22-1 Patrick Gelsinger, Intel CTO

Patrick P. Gelsinger
Senior Vice President
General Manager, Digital Enterprise Group
Intel Corporation

Pat Gelsinger is senior vice president and general manager of Intel Corporation's Digital Enterprise Group. Gelsinger joined Intel in 1979. As CTO, he coordinated with Intel's longer-term research efforts and helped ensure consistency from Intel's emerging computing, networking and communications products and technologies.

Before his appointment as the company's first CTO, Gelsinger was the chief technology officer of the Intel Architecture Group. Previously, Gelsinger led the Desktop Products Group. From 1992 to 1996, Gelsinger was instrumental in defining and delivering the Intel ProShare video conferencing and Internet communications product line. Prior to 1992, he was general manager of the division responsible for the Pentium Pro, IntelDX2 and Intel486 microprocessor families.

Gelsinger holds six patents and six applications in the areas of VLSI design, computer architecture and communications. He has more than 20 publications in these technical fields, including "Programming the 80386," published in 1987 by Sybex Inc. He has received numerous Intel and industry recognition awards, and his promotion to group vice president at age 32 made him the youngest vice president in the history of the company.

Gelsinger received an associate's degree from Lincoln Technical Institute in 1979, a bachelor's degree from Santa Clara University in 1983, Magna Cum Laude, and a master's degree from Stanford University in 1985. All degrees are in electrical engineering. Gelsinger is married and the father of four children.

Figure 22-1 Dr. Bernie Meyerson, IBM CTO

Bernard S. Meyerson, Ph.D.
IBM Fellow, Vice President and Chief Technologist
IBM Systems and Technology Group

Dr. Meyerson is Chief Technologist and a Vice President of the Technology side of IBM's Systems and Technology Group. Dr. Meyerson joined IBM Research as a Staff member in 1980, leading the development of silicon:germanium and other high performance mixed signal technologies over a period of ten years. He subsequently has led several large organizations within IBM focused on the development of communications technology, encompassing business applications spanning the range from pervasive wireless enablement (802.11x) to high end data transport. Most recently, Dr. Meyerson assumed responsibility for Microelectronics Technology Development as head of the SRDC (Semiconductor Research and Development Center).

In 1992, Dr. Meyerson was designated as an IBM Fellow, which is IBM's highest technical honor. He is also a Fellow of the American Physical Society and the IEEE. Throughout the years, Dr. Meyerson has received several awards for his works including: Materials Research Society Medal, the Electrochemical Society Electronics Division Award, 1999 IEEE Ernst Weber Award for the body of work culminating in the commercialization of Si-Ge-based communications technology, and the IEEE Electron Devices Society J. J. Ebers Award.

He was cited as Inventor of the Year in 1997 by the NY State Legislature, and was honored as the 1999 United States Distinguished Inventor of the Year by the US Patent and Trademark office. In 2002, he was elected to the National Academy of Engineering.

Chapter 23

Holistic Design

Future Chip Innovation

In his 2005 DAC keynote and again in my interview with him, Dr. Bernie Meyerson stressed what he calls holistic design as a key for chip innovation in the future. Meyerson is quick to point out that holistic design is a solution, not a technology.

The Interview

Right after we toured the IBM East Fiskill NY plant, Joe McDonald, my publisher and I had lunch with Dr. Bernie Meyerson, the same IBM Vice President and Chief Technology Officer to which I have been referring regarding his DAC Keynote speech and his mini battle with Intel's Patrick Gelsinger. Other than the 10 minutes to get to the IBM cafeteria, which included Bernie having to open the windows of his "Vet" to relieve it of some July heat, I had about two hours for my interview with Dr. Meyerson. From my perspective, he does not look like he dates Martians, but he does look like he likes to have a lot of fun and he sure has a lot of fun in his job.

In the interview, we discussed many of the issues of classical scaling, Moore's Law, IBM's heat solutions, and many of IBM's innovations. Meyerson had already run through these notions in his DAC keynote and had them on the tip of his tongue in our interview. In a nutshell, Dr. Meyerson believes that innovation is the key to performance boosts at a system level and innovation in chip design is just one part of the whole.

Think Outside the Chip

Several years ago Meyerson introduced a notion to the IBM chip making process that was fairly unique for the semiconductor industry. He calls it the holistic approach to design. Though Meyerson's thousand or so engineers in the NY facility are primarily focused on designing IBM's processor chips and working with IBM's partners in Power.org and elsewhere, under Meyerson, they are literally challenged to think outside the chip.

So that we are all on the same page, I am including a definition of holistic which fits Meyerson's thinking. The word holistic is an adjective that emphasizes the importance of the whole and the interdependence of its parts. So, it would follow that holistic theory would be that the parts of any whole should be considered in relation to the whole, and that the whole is often greater than the sum of its parts.

Without getting into medicine or various controversial holistic philosophies, we can simply offer that holistic design suggests that as parts are designed, the overall design of the system or entity needs to be considered. Additionally, as changes in technology occur which affect parts, such as the potential for better parts or new parts or components of parts, as these are brought forth it calls for a re-evaluation of the whole entity as it exists to see that a different arrangement of parts or functions on the parts will produce an even better entity.

Chip and System Level Integration

Meyerson's quick term for the purpose of holistic design is that it is all about value creation- the materials, devices circuits, cores, chips, system assets, system architecture, system software, application software etc. It provides the most effective means to optimize the value of IT offerings to the end user because all parts, including the user part are considered. From the chip designer's perspective, it's now about chip and system level integration using inside and outside resources to satisfy the requirements of the ultimate user.

Before we explain this further, let's take a look at a specific chip that was built from holistic design to be the best that it could be – The Cell Processor. This is a great example of application specific computing in the game console market. It is the ultimate in which holistic design can take a developer / engineer. Its purpose is to enhance the experience at a Sony Playstation 3 and its design reflects its intended purpose. None of the three partners, IBM, Sony, or Toshiba was looking to build a general purpose chip and so, all aspects of the intended use were considered and as much as possible was built on the chip along with the interfaces to the controlling software and the physical interfaces to the ultimate game cartridges.

The moral of holistic design is to start by thinking what you want to do at the end and then do it. If it's a game console chip, look at the tools you have and what you need and do it. That is the future. Innovation will drive future technology performance. Blind scaling has run some companies into a thermal wall. Meyerson likes to say that processor design and metrics have changed irrevocably so there is no old recipe book to fall back on. Innovation through holistic design will carry the future.

Where Have All the Gigahertz Gone?

As noted in Chapter 22, in his DAC presentation, Dr. Meyerson postured that "gigahertz are gone." They have been replaced by application specific metrics achievable hthrough innovation. Design for processor frequency is no longer an issue because it no longer makes sense – even for Intel. So, system solutions can now be optimized via holistic design which will ultimately dominate progress in information technology. Application specific computing such as the Cell chip revolutionizes computation costs and a fiscal reality has driven and is driving the semiconductor industry towards technological consolidation. Not only was a holistic design used for the creation of Cell but Sony did not do it alone. Sony and Toshiba partnered with IBM to assure that the mission was accomplished right.

Pure Research and Computer Systems

IBM is fairly unique in the chip industry in many ways. **First of all**, the company not only conducts chip engineering and scientific research, it also conducts pure research into all sorts of things. One of the key areas of pure research for semiconductors is materials. But, in addition to materials, IBM spends a great deal of its resources on researching techniques and approaches that will be necessary ten years or more down the road for the chip industry to continue to provide performance gains. Because research is such a key to success in the semiconductor industry, in Chapter 24 we discuss research at AMD, Intel and IBM and its fruits in greater detail.

Secondly, unlike Intel and AMD, in addition to being a leader in chip technology, IBM is also a traditional computer system company. Big Blue has made computer systems, now called servers since the early 1950's. So, IBM understand the role of the processor, memory, storage, and all the various interconnects and network channels necessary to build a complete computer system. To build chips today that enhance the function and performance of the server as well as the people who design application systems, those who program, and those who use the server, the chip designer's job is more difficult. For example, he or she must communicate and operate at many more levels. In the past, for example, for some companies, clock frequency was the only thing that appeared important. Chip designers today must be good computer scientists, good engineers, and good communicators.

Design Input from Many Sources

So, IBM has two things that its competitors do not have as input to its holistic chip design process – 60 years of pure research and a 50 year system orientation. IBM has the results of its pure research, and as a corollary, it has cross licensing agreements with most companies for research results and technology that IBM has not invented. IBM also has all the inputs that it can get from its research and technology divisions as well as its four server divisions and its software division. Unquestionably, IBM has more input to its chip design process than any of its competitors. Quite frankly, it shows in Big Blue's products today as well as in its roadmap for the future. On top of this valuable input, in its site in East Fishkill, IBM has on-site fifty top engineers each from companies such as Sony, Chartered Semiconductor, AMD, and others. The interactions with these outside experts provide another source of input in IBM's holistic design.

Virtualization

One of Meyerson's pet areas as an objective of holistic design is the notion of virtualization. According to Dr. Meyerson, there are various types of workloads that arrive at a system, each demanding different combinations of resources. He cites as an example, the verification of a five digit pin number. Now, in the instant that this workload is processing in a dual processor design, one processor is tied up for that short duration working on just a small stream of instructions, yet it can't go anyplace else. If we offer that the processor has the hardware to run two threads simultaneously, then effectively, this small task is tying up one whole thread or twenty-five percent of the system until it is finished and has posted its results. This type of workload does not make efficient use of processor threads and other resources that are actually powerful enough to guide a rocket to the moon and back

So, what is the solution? Well, having four threads on one chip certainly helps overall but with varying types of workloads such that at one time a whole processor may be needed and at another time just a minute part of a processor may be needed, there needs to be a better way. For those programs that are computational oriented and those that perform few computations but drive huge amounts of date through the system, there needs to be a better way.

Traditional IBM uniprocessor chips have been built for many years for multi-user, multitasking environments in which the operating system manages the workload through things like job control and task switching and hardware interrupts etc. The operating systems on modern day symmetric multiprocessors (IBM's iSeries and pSeries each use as many as 64 POWER processors in parallel) still perform many of these functions off the chip but IBM has actually built a better way and it is getting better with each chip release.

Holistic Design for Virtualization

In holistic design then, the engineer gets to consider how the chip is actually going to be used by memory, by storage, by software, etc. and he gets to think about ways to have the chip perform tasks that were once performed by the operating system. Considering that IBM writes its own operating systems and has major input to the Linux movement, an IBM chip designer may build things into the chip in total or in part. For example, if there were not enough chip real estate to provide a means of extended parallelism to be able to execute more tasks concurrently on a chip, today's chip engineer can take another approach. With a deep computer science background to accompany their engineering skills, IBM's chip engineers have the knowledge to design an approach that may perform some of the work on the chip and some of the work in the operating system. They may even choose to use an application program for other parts of the system.

A holistic approach takes the whole end product and its uses into consideration. One of the enhancements that have come about from this approach is the concept of micro-partitioning. Micro-partitioning permits the thread capabilities of the system to be expanded substantially. Even before the POWER4 generation of chips, IBM had implemented partitioning with its iSeries machines. With POWER4 and POWER5, IBM has taken the concept and simplified its use. Instead of having a small job hold up a thread as discussed above, with partitioning, each thread can be divided into ten micro-partitions so that work can then be assigned to a micro-partition and the other micro-partitions can be available to process the next set of work that comes in.

Dynamic Work Assignment

Consider this. The operating system or something even better, a hypervisor, examines the work request, and based on the work, it chooses to assign the job to a micro-partition that operates as a virtual machine. For argument purposes the virtual machine takes 1/10th of a thread or 1/20th of the processor. So, when this PIN verification work is assigned as discussed above, it does not hold up the whole thread, it takes just 1/10 of the thread's resources and overall just 5% of the CPU. If a big job comes in next that could use the whole CPU, the operating system and the chip together give it as much as it needs up to 95% of the system and if another job comes in after that, the OS and chip apportion resources from the big job so that the new job and the little job do not have to wait. You get results like this only with a holistic approach to design.

Hint Wikipedia defines hypervisor very simply as follows: "In general terms, a virtual machine in computer science is software that creates an environment between the computer platform and the end user in which the end user can operate software." A hypervisor can control partitions and can support multiple operating systems – one each in each partition.

The Chip is Just one Virtualized Puzzle Piece

The notion of virtualization provides for more than processor apportionment. There are lots of other resources that come into play, especially in larger systems. Virtualization presents the physical resources to the "users' with the same interfaces / functions, but with different attributes. A virtualized object may be a subset or a part of a physical resource or it may be one interface to multiple physical resources

So, a system creates virtual resources and then "maps" them to physical resources with software and / or firmware. These are real hardware components with architectural interfaces / functions that permit them to appear in multiple areas to multiple users. The types of resources again include a centralized or distributed CPU, memory, storage, servers, etc. The notion of virtualization separates the presentation of resources to users from actual physical resources so a specific user may actually believe that they are the only one with the resource.

Hint: The notion of a distributed CPU brings connotations of grid computing. If the CPU is virtualized, it really can exist in another computer system in another country for that matter and it is reached via an optimized network.

Design for virtualization therefore includes the aggregation of pools of resources to users as virtual resources. Thus you see virtual storage, virtual partitions, virtual LANs and other virtual entities as if they are real. The effect on the user is positive. The effect on the system is that it is far better utilized.

For the most part the notion of virtualization through the use of micro-partitioning as described above is very real and soon grid computing will be just as real. As a system manufacturer, IBM knows how to deal with all of the issues associated with both. The company is already a champion of supercomputing in which the notion of resource virtualization has always been prevalent.

IBM Leads in Holistic Design

At the chip level, today's POWER processors have partitioning and virtualization hardware built on the chip. They support virtualization by dynamically reconfiguring the partitions based on the workload definitions provided by the hypervisor and/or the operating system. Software and hardware are each aware of each other-- with each helping each other with the workload-- that's what it is all about. That is the payback of holistic design. Since no other chip maker in the industry, including Intel has all the inputs of IBM nor can any other chip maker design at all these different levels with the same quality of input, holistic design gives IBM a phenomenal advantage for now and for the future.

Chapter 24

The Fruits of Research

AMD Research Budget – Less than $5 Billion

Go ahead, I dare you. Try to find out what AMD's research budget is. IBM's R&D comes to about 5% of its gross margin and Intel's R&D is about 10% of its gross. Intel is about 40% the size of IBM. AMD is about 12.5% the size of Intel. So, I would guess that AMD's research budget must be substantial because they do so much better than Intel but they are 87½ % smaller. So, on my own, I figured they had about a 20% research budget or about $1Billion.

When I first wrote the paragraph above, I had not been able to find anything about AMD's R&D. Then, from my many accounting courses as a data processing major in college, I realized that was not going to fly. Nobody would accept that this information was unavailable. In fact, the more I thought about it, I even couldn't accept that it was unavailable. But, it was hard to find. Then I recalled the annual report that all companies with stockholders must produce. I looked up AMD's and voila, there it was in black and white, a sum total of $934,574.00 given in thousands. So, now I feel pretty good about myself. AMD's R&D is just under $1Billion per year.

How Does AMD Spend Its Research?

No matter how you look at it, AMD is getting a good deal from its research dollars. From the development of its high speed Athlon at the turn of the century to breaking the 1GHz barrier before Intel, to beating Intel in the x86 space in 2003 with the groundbreaking

AMD64 technology and the Opteron processor, without even mentioning its industry leading MirrorBit flash memory technology, AMD has done quite well. Add its entrée to dual core in 2005 and the R&D folks at AMD get an "A+" for a great job. As a professor, myself, I know how hard an A+ is to achieve.

Not only did AMD bring 64-bit computing to the masses in 2003, it was with no penalty. The company engineers and scientists designed an extension to the x86 architecture that included about ten additional instructions to support a 64-bit architecture. Not only was it an easy jump to 64-bits but AMD also included 32-bit hardware on the chip so that old 32-bit Windows programs and old 32-bit Linux program would run untouched on the new 64-bit chip. Knowing that was not enough, AMD engineers worked with Microsoft so that its operating systems could use the additional 64-bit instructions and could switch dynamically from 32-bit to 64-bit computing on the fly. Moreover, AMD along with some R&D help from IBM brought its Opteron's performance to nothing short of outstanding, beating everything from Intel by a substantial margin.

While AMD was splitting its R&D so that it could make major innovations in its MirrorBit memory technology, the company was also developing an Athlon version and dual core versions of its 64-bit processors. In just a few years and with a few research dollars, AMD was able to outclass Intel in its cash cow area – x86 technology. At the Microsoft Tech Ed 2005 Conference, the Dual-Core AMD Opteron processor took "Best of TechEd 2005" honors as conferred by Windows IT Pro and SQL Server Magazine. Additionally, Network Magazine named the Dual-Core AMD Opteron processor "Breakthrough Product" for 2005 at the Network Magazine Innovation Awards. Somehow, AMD's research dollars had paid off.

Hey Big Spender

Meanwhile there was Intel out there with lots more R&D to spend? And, what did the big chipmaker do with all its R&D money? Quite frankly, I don't know. Lots of industry analysts don't know. Intel engineers or Intel management or both missed the problem with classical scaling completely and produced a product that could serve well as a portable heater for the wintertime rather than a product that would drive system performance through the roof.

From my examination of Intel press releases and industry articles over a three to four year period, Intel never really admitted that it had made a mistake. Hey, with all that R&D and no results, the company made a mistake so big that its major competitor, AMD, spending its own pennies on its own research and buying what it couldn't do itself from IBM, created a machine that outperformed the best that Intel could build.

While this was going on, Intel's engineers were not looking at the future; they were frantically trying to solve a heat problem that would not be solved. All somebody has to do is listen to IBM's Bernie Meyerson's 2005 DAC Keynote speech and they would understand intrinsically that classical scaling doesn't work any more. So, why were Intel's engineers trying to fix a problem that could not be solved, for so long before they cancelled the project? Eventually Intel told the world that their efforts were unachievable -- 4GHz was unachievable. But, by the time they told the world, they had already invested over three years of R&D trying to solve the problem and they had given the public a line of half truths that made their predicament even worse.

Intel Moves Forward

Of course Intel's engineers were also working on Itanium, and maybe that big trough to nowhere is where a lot of R&D dollars went. During the time that the giant chipmaker was working on the heat problem, it did a few good things. They weren't necessarily groundbreaking, but they were good. Rather than try to convince Microsoft to create a new version of Windows for an Intel-unique version of AMD64, which would have been a PR disaster, Intel ate some crow and cloned AMD64's architecture and created Xeon 64-bit chips in just about a year that were almost as powerful as AMD's. Intel also carried the 64-32-bit trick into its other x86 chip designs.

Finally, Intel management made a commitment to parallelism and the company began in earnest to use innovation, rather than clock frequency to achieve performance boosts. In 2005, Intel moved its target for the future to the AMD64 style of 64-bit x86 computing along with dual core processors.

These were all good things but one would expect lots more than "me too" from such a huge research budget.

Intel Research Budget - $4 Billion

Intel spends about $4 billion each year in R&D but its R&D philosophy is very different than that of IBM's traditional pure research approach. Of course even IBM isn't really like IBM was regarding R&D. When you think of who really practices real R&D in the US in the high-tech sector, at one time you would come up with names like Xerox Palo Alto Research Center (PARC), where GUI was invented, MIT Lincoln Lab, AT&T Bell Labs (Lucent Technologies) and of course IBM's Thomas J. Watson Research Center.

Those were the good old days for scientists and engineers involved in basic, fundamental research. Life in high tech research is no longer done country club style as corporations have begun to demand more and more practical output from the smart folks who once got paid just to sit around and think.

Today, Bell Labs, Watson Research and Xerox PARC are nothing like their former selves. They were downsized by the realities of profit making organizations that now put the highest premium on short-term product goals rather than long-term research results. Above all purposes in a corporation, survival is #1 and profit is #2. These large corporations changed their ways because they believed that they could still produce high quality products while being more conservative on pure research spending. Moreover, if an effort such as, "why is there air," has no chance of a realistic payback, the study in today's world, as important as it may be to basic research, is cancelled. The dollars are shifted to something that can bear fruit much sooner.

IBM Research & Intel Research

There is a major difference even in today's research marginalized world between how Intel sees research and how IBM sees research. For all of its spending in its many "Lablets," with the directed purpose of finding solutions that can become products, Intel's track record at the product level and the product improvement level in the last five years or so is not very impressive. So, where did all those dollars go?

CTO Patrick Gelsinger is the guy who makes the decisions as to how Intel spends its R&D dollars. Gelsinger has set Intel up with lots of little R&D facilities scattered all over the world. He acknowledges that his company is nothing like the model of a Bell Labs or the old IBM or Xerox PARC and he thinks that is good.

Intel has over 75 R&D facilities around the globe, each containing a small number of researchers often focused on specific areas of research and/or development. Intel in Israel is doing so well for the company for example, that it now has about the same number of employees that Intel employs in Silicon Valley. Intel believes that Israel is loaded with technological talent and the giant chipmaker does its best to capitalize on that. So also with its many other labs and "lablets."

IBM, on the other hand has about 8 major Labs. The company spends about 6 Billion dollars on research each year, which is more than Intel but less than the IBM of yesteryear. Though IBM cut its pure research percentage of that budget substantially in the mid 1990s, as it was trying to survive, seeing the need for pure research, especially in materials, that portion has grown back to about its original size.

The University Connection

Intel partners with universities for basic research as does IBM. In fact, Intel and IBM have arrangements with many of the same universities for different projects. Intel's purpose is to find the top researchers in the field at universities and engage them in pure research projects that will benefit Intel. The company's approach is not to try to hire them but, instead Intel throws them "seed" money to help control projects of interest both to the researcher and to Intel.

Of course that does not mean that Intel won't hire somebody if they really want the person. Ironically, Craig Barret, Intel's past CEO and current chairman is an exception to the no-hire rule. In the early 1970s, Barret was knocking them dead at Stanford, with a PhD in materials science. Intel came along one summer and asked for a student to help sift through a problem with the ceramic packaging around one of the company's new chips. For the heck of it, Barret offered himself. The to-be CEO quickly solved the problem and was sold on Intel. Since then, from Intel's perspective it's been "Barrett Inside."

Intel works very closely with leading colleges and universities around the world on research projects that the company expects to bear fruit as far out as the 2020 time-frame. It is these relationships that allow Intel to offload the vast amounts of early research to the universities and then the company selects the most promising techniques.

One of the biggest differences with Intel and IBM R&D is that Intel has never fully engaged in pure research to the extent of IBM. That does not mean that Intel does not spend lots of money. Intel spends tons of money in R&D, just not the same way that IBM spends its money. Intel's dollars are spent more on the development side to assure that it has top flight manufacturing plants ready to produce its latest product.

Dollars for Problems

The giant chipmaker also spends lots of money on problem areas and potential problem areas. Sometimes the problems that need development solutions are several years off in the future, but recently the problems have been such that it would have been good for Intel if they had been solved a few years sooner. Moreover, Intel may very well find out that in order to solve the problems of the day, it would be good to have had some pure research solutions in the pipeline. The inordinate amount of time and effort the company expended to determine that it could not get beyond the heat barrier is a prime example of how early research would have helped help.

Intel looks at "pure research" as a problem that typically takes about five years to solve. When Intel's researchers grasp the problem and know how to solve it, the company moves the project into its "advanced development" phase which may be two to three years from a solution. That does not leave a lot of room for making a big mistake such as defining and announcing a chip architecture that cannot be manufactured.

The Intel goal from its research is products, products, and more products. It's hard to believe but the Santa Clara corporation claims that it does not expect all research inventions to eventually wear an Intel logo. However, the giant chipmaker has had some issues in coming up with agreements at Universities as the academics of today are also looking for a share of the IP benefits. To counter this trend, Intel has begun to partner with more non US academic institutions that are not as eager to grab a share of the spoils.

Gelsinger's Intel sees the lab not only as a development place but as a major testing place. In fact, Intel offers that it prefers that its labs fail in bringing technology to the market, rather than have the product fail once it is released. Unfortunately, over the last several years this goal has not been attained in some major areas.

Again, the bottom line of Intel's research is tilted towards products. Unlike IBM, a company that still takes pride in the scholarship of its researchers and shares the recognition of prizes, such as the Nobel Prize, Intel has a different philosophy. The company really does not want to know how many Nobel laureates it has produced but how much research they do and how effectively the company can translate that research into products. .

Intel Wants the IP

In cases where Intel needs to partner with a university, that often means negotiating intellectual-property agreements. Apparently U.S. universities negotiate more stringent IP deals, so Intel has focused more of its research efforts overseas, where it can work out more favorable IP ownership arrangements. As far as patents are concerned, Intel is proud of its patent record but the company believes that 1000 patents a year are more than enough.

Win through Standards Committees

Rather than concentrating just on R&D, Intel has found that committees have power over the standards that control a particular technology, which can eventually lead to control of the market. So, Intel is always ready to join a standards committee. Their philosophy is that it doesn't help much to create a defining technology if the standards committees select a different approach. Controlling the standards committees is one key to Intel's success.

How Did Intel Get In Trouble?

The more I examine Intel, the more I see a company that lives for its next sales opportunity. The company seems to be more proud of its immediate manufacturing abilities than research for its products. The giant chipmaker's biggest R & D failing over the past five years is that it unknowingly kept increasing the clock frequency of its chips until it could not achieve the performance objectives that it had announced.

I have reported on this several times in this book but it really is a big puzzlement for me. Did Intel have the R&D in place that was necessary to gain performance benefits through innovation? Did management choose to go with the classical scaling approach to gain the "sure thing" without concern for research? Either there was no research dictating that the sure thing was no longer a sure thing or despite the research, Intel management believed that its developers would solve the heat problems and the scaling problems associated with high frequency processors in time for a successful product launch.

It is very hard to believe that the world's largest chipmaker, with the largest R&D budget for chip development in the world did not have the necessary research input in its pipeline to solve one of the biggest problems of all time at Intel. That problem arrived on time somewhere between the 130-nanometer and 90-nanometer switchover.

Where's the Beef?

So, it looks like results from Intel's research in its biggest game – x86 processors has been absent from the foray. A $4 Billion research budget that does not understand the limits of classical scaling and the problems brought forth by overheating chips just does not compute for me. Moore's Law or not, I expect that any company that spends or says it spends that much on research ought to have a few solutions in the bank. At a minimum, a company with a $4 Billion research budget should know that it has no solution and thus spare itself the embarrassment of announcing capabilities that it cannot produce.

IBM Research Spawns Innovation

"Innovation sprouts from the fertile ground of inquiry." In IBM Research, the company's foundation is its wide array of scientific disciplines, award-winning researchers and global labs. And, over the years, IBM has expanded its ground to include challenging ideas from colleagues in academic and government research centers as well as from many of its clients and partners. IBM maximizes its return on its research in many healthy and visible ways.

Intel became a successful corporation because of its own innovations. Intel became a large corporation because it had one big client in 1981 that took a fledgling company and helped make it a superstar. Since its inception, the giant chipmaker has been immersed in its desire to bring products quickly to market while retaining any and all IP rights in its research efforts. Intel is a cunning competitor and lately the company seems to be working much harder to assure its market dominance than in building innovation into its products. Maybe AMD has a case but that is an issue for another day.

I have been very critical of IBM in other books that I have written for the opportunities that Big Blue has squandered over the years. If IBM capitalized on the results of its research, Big Blue would be a four hundred billion dollar company and, as a stockholder, I would be in the Bahamas right now. Over the years, IBM was given a good hand from its research, but it did not play its cards well. Lots of companies made lots of money on IBM inventions.

Intel in comparison to IBM performs very little pure research. IBM has always been the leader in pure research. In fact, Intel, Microsoft, Sun, Oracle, and others have benefited, perhaps even more than IBM from Big Blue's basic research without having to pay the price of that research themselves. The IBM that serves the semiconductor industry today, however, is a far different IBM. If Intel has not noticed the new IBM, it had better.

IBM Research Environment

IBM now fosters a more open development environment, especially in the semiconductor area. A simple walk through IBM's East Fishkill plant and it is obvious from the languages spoken and the diverse mix of people that it is not just IBM people who are engaging in research. IBM partners from companies all over the world have as many as fifty engineers each working with IBM engineers on specific projects in which all will share some of the IP.

From years of doing research and from years of being in the forefront of technology, the new IBM seems to have gotten it right this time. Intel, for all its billions spent on research, has little to show.

The purely technical disciplines in which IBM conducts research today include the following:

- ✓ Chemistry
- ✓ Computer Science
- ✓ Electrical Engineering
- ✓ Materials Science
- ✓ Mathematical Sciences
- ✓ Physics
- ✓ Services Sciences
- ✓ Management and Engineering

In addition to these, there a number of cross-disciplines in which IBM engages. These include the following:

- ✓ Communications Technology
- ✓ Deep Computing
- ✓ Display Technology
- ✓ E-commerce
- ✓ Personal Systems
- ✓ Semiconductor Technology
- ✓ Storage
- ✓ Server & Embedded Systems

IBM's Research Happenings

IBM is far ahead of Intel in almost everything that counts from its research to its on-chip technology to the system and software products that flow from its research. Judging from results, there is something systemically wrong with Intel's research model and its execution. It is certainly beyond my ability to offer a comprehensive solution to Intel, but I would suggest they look at what IBM is doing. Intel should consider swallowing its pride and getting help from IBM in both management and engineering services. The alternative is that the world's biggest chipmaker will continue to be in the embarrassing situation of having to clone technology from AMD, a company that chooses to benefit from its IBM relationship.

IBM's research community and its accomplishments are so substantial, they are always in the news. Two recent news articles help to demonstrate Big Blue's typical groundbreaking research accomplishments. Both stories show how IBM's research prowess has come to fruition in the form of some exciting products and industry awards.

Reflecting IBM's major research investments, its innovative PowerPC chips, its clustering, networking, symmetric multiprocessing and simultaneous multi-threading, as well IBM's long-time leadership in designing whole computer systems, Big Blue is now the undisputed leader of supercomputers.

IBM and Supercomputers

IBM research in the above areas has paid off big time for the corporation in tangible results. On June 22, 2005, the world's foremost supercomputer ranking authority, Top500, (www.top500.com) named an IBM supercomputing system as the most powerful supercomputer in the world. IBM's BlueGene/L topped the list with a sustained performance of 136.8 Teraflops, or trillions of floating point calculations per second. The system developed with IBM's primary partner, the Department of Energy's National Nuclear Security Administration, is partially installed at Lawrence Livermore National Laboratories in California and is not yet finished growing in power. It is planned to grow to a 360 Teraflop BlueGene/L supercomputer when completed in a few months.

IBM Research Division is especially proud since this is the first time in history that a single vendor has more than 51% of the total number of supercomputer systems on the list. IBM is the leading provider of both installed supercomputing systems with 259 systems as well as total aggregate supercomputing power, with a record total 976 Teraflops. IBM has 6 of the systems in the Top 10, including MareNostrum, Europe's most powerful supercomputer which is powered by IBM's PowerPC Microprocessor and eServer BladeCenter JS20 - the only supercomputer based on blade server technology to ever be ranked in the global top 5. According to numbers compiled by the TOP500 List of Supercomputers, IBM is also the overwhelming leader in global supercomputing with 57.9 percent of the total processing power, compared to its closest rival, HP, which comes in at 13.3%.

Adding to IBM's plaudits, if we were just to include IBM's BlueGene install base as reported by Top500, it would be the equivalent to the total number of Cray systems and about 60% of all SGI systems on the list. Considering that Cray is the eternal leader in supercomputing and has concentrated mainly on supercomputing as its main product line, this is quite an accomplishment for IBM.

Today's single chip microprocessors are taking on more and more of the personalities and capabilities of supercomputers. It sure helps a company that is working to bring supercomputing to a chip to also be the top performer in the supercomputing marketplace. This gives IBM a major advantage for the future as Intel, for example, has not engaged in the supercomputing race. Intel chips are used by others in supercomputers- even IBM supercomputers – but not BlueGene/L. Additionally, Intel is not a supercomputer manufacturer, it is a chipmaker. Intel is not a system maker, it is a chipmaker. Intel is not a software company, it is a chip manufacturer. In the future, that just might not be good enough.

Invention of DRAM (Chip Memory)

In an earlier news event, one of IBM's top researchers, Dr. Robert Dennard, on April 22, 2005 was awarded the Lemelson-MIT Lifetime Achievement Award for his invention of dynamic random access memory (DRAM). The award honors a "remarkable individual for his or her lifelong commitment to improving society through invention" and cites DRAM as a "breakthrough that transformed the microelectronics industry in the early 1970s and remains the most popular form of computer memory today."

During his early work on the technology, Dennard referred to his invention as "one transistor memory cell." Once it was in use in the industry, it was renamed DRAM. "I don't know if people really understand why it's called dynamic. It's because it stores the charge only temporarily and that charge leaks off within a fraction of a second because the transistor it's connected to is leaking. And the charge has to be read out before it leaks away and written back in – called refreshing – and that's how the term 'dynamic' came to be used."

He relates an amusing anecdote about the final product name, from the laudatory address given when he won the Aachener and Münchener award for technology and applied natural sciences in 2001. "The speaker quipped that my contribution was having the courage to call something that only retains its information for a fraction of a second *memory*."

Collaboration

Dennard believes in the importance of collaboration, and he points to his extensive list of publications as one example of the rich collaborative nature of his work. "Ghavam Shahidi, Matt Wordeman, Bijan Davari and many others – I'm so lucky to work with so many talented people."

"Collaboration with other IBMers has been invaluable," he continues. "You can't really work in isolation. You have to be in touch with what the real problems are – and they are complex, needing a wide range of disciplines to solve them-- Physics, materials engineering, processing, lithography, semiconductor device understanding, circuit design, and, of course, financial considerations. You have to collaborate with various people who have all this knowledge. It's one of the most exciting things about working here. I don't plan my day – I just walk around the lab and see who I meet. Because that may be the person I'm looking for to answer a question."

Of course in the modern era, there are even more scientists with whom to collaborate. In concert with many of IBM's partnerships in research activities, Bob Dennard and the team of IBM inventors get to work with engineers and scientists from other companies around the globe in a common effort to conquer the limits of technology.

Gordom Moore Lauds Dennard

"It is truly unusual in a field moving as rapidly as semiconductor technology to have an invention endure over three decades, but there is still no technology on the horizon to replace the single-transistor DRAM cell," Gordon Moore, co-founder and chairman emeritus at Intel, wrote in his recommendation letter for Robert Dennard.

Before Dennard created the DRAM cell at IBM in 1967, technology companies had been using magnetic core memory and were having a difficult time using it with faster processors. DRAM was much smaller, required less power and cost less and ultimately was much faster than its bulky predecessor.

IBM introduced the first commercial DRAM chips in the 1970s. Today's version of those chips can hold up to a billion bits of information and provide the RAM function in all sizes of computers from hand-held to supercomputers. DRAM is also the memory of choice in pervasive applications from communications network switches to digital cameras.

More Inventions

DRAM was not the end of Dennard's accomplishments. In addition, Dennard also received the award for a significant breakthrough in integrated circuits (IC). In the early 1970's Dennard and his team at IBM took on the challenge of whether bipolar-transistor-based ICs should be the future building blocks for high speed electronics. . Dennard did not think so. He preferred the metal-oxide field effect transistor (MOSFET) since, from his findings, it offered a better alternative because of its amenability to large-scale integration with high density and good yield.

To prove his point, he and his IBM team developed a scaling theory, and demonstrated that if all dimensions of a MOSFET device were reduced simultaneously, along with other factors, they could continue to make smaller and smaller devices that performed better, required less power, were denser and less expensive. IBM and the semiconductor industry owe an awful lot to Bob Dennard, an IBM researcher whose research paid off big-time.

"Dennard's development of scaling theory has been a driving force in microelectronics," wrote Nicholas Donofrio, IBM senior vice president of technology and manufacturing, in his recommendation letter.

Dennard's paper devoted to this theory is universally referenced as a guide to designing devices down to submicron dimensions. Throughout his career, his work has resulted in 35 patents, nearly 90 published technical papers and numerous awards, including the National Medal of Technology.

It can be argued that the most relevant inventions in the modern computing era are ☞ the integrated circuit, ☞ MOSFET, ☞ the one-transistor DRAM cell and ☞the magnetic disk. Considering Dennard invented the DRAM cell and his work in scaling helped MOSFET become more widely adopted, this was a significant award, indeed. Considering that IBM also invented the magnetic disk, IBM has two out of the four major inventions of all time.

IBM Knows Research

IBM spends about $6 billion a year to support over 3,000 researchers in eight labs around the world, including the Thomas J. Watson Research Center in Westchester County, NY. Admittedly, that's not much more than Intel spends but IBM has lots more to show for it.

IBM's engineers and scientists have lots to their credit. They invented the hard disk drive, the scanning tunneling microscope — which offered the first-ever look at individual atoms — and IBM engineers discovered high-temperature superconductivity. The latter two achievements were rewarded with Nobel Prizes in physics.

In fact, five IBM scientists have been awarded Nobel Prizes for their efforts. Leo Esaki was awarded the Nobel Prize in Physics in 1973 for his discovery of tunneling in semiconductors. Heinrich Rohrer, along with his colleague, Gerd K. Binnig, was awarded the Nobel Prize in Physics in 1986 for his work in scanning tunneling microscopy. J. Georg Bednorz, along with his colleague, K. Alexander Mueller, was awarded the Nobel Prize in Physics in 1987 for his discovery of high-temperature superconductivity in a new class of materials.

Research from 1945

IBM founded its research arm in 1945. Its purpose was to explore ideas. When I joined IBM in 1969, IBM had an Advanced Systems Development Division (ASDD) and a Systems Development Division (SDD). We joked about the ASDD guys wearing fur coats and sneakers and beards and t-shirts while the IBM standard uniform at the time was a suit and a white shirt. Yet, nobody dared say anything about the researchers and their garb. They worked on pure research that was ten or more years out. The SDD folks took the results of their work and merged it with business requirements and in a five year horizon built things that may or may not ever be marketed. The word internally was that less than 10% of the projects got out the door as products, but it was always good to have that research tucked away.

IBM's research people invented a lot of things that made the company successful and some that made others successful first. For example,

IBM's John Cocke invented RISC (made Sun very successful) and Tedd Codd invented relational database (made Oracle very successful)

Changes in Research Style

In 1976 or so, IBM changed its research structure as it created a number of centers and the people in the ASDD, for example were merged into other research areas within the company. Things changed again during the last of the Akers years as IBM realized that getting ideas out of the labs and into the market faster meant more opportunities. To do that, however, there were lots of barriers to be torn down between research and development.

At the time of the big change, IBM employed about 2,500 pure researchers who were accustomed to doing basically whatever they felt like doing. They would spend days — and often nights — considering the great scientific questions of the day. They had no concern at all to turning any of those thoughts into products. And they did not come to IBM to be part of a development team

Some were swayed by the success of their peers who did make the leap. For instance, Dr. Bernie Meyerson and his research team pioneered the super fast silicon germanium transistors in the late 1980s. Initially these researchers were not concerned about turning their technology into microchips. However, they were quick to complain when the development team took the baton at IBM's East Fishkill plant. To make the chips more manufacturable, the development team changed the process that the researchers had defined. The research scientists accused the developers of ruining the product. In a great act of defiance, the research team started its own competing development effort.

So, for a while, there were two teams making the transistors. In the end, though, researchers and developers chose to work together. Of course some chose to move to more pure research areas such as academia and a few others left but most stayed and for the most part, they actually liked their jobs more. They were able to stay with their inventions to the end without the compromise of a handoff.

Unexpected Learning

Scientists and engineers who went through the change in fact have offered that they never learned more than when they started getting involved with product people. Though the engineers knew lots more than the product people and thought the product folks were basically ignorant and shortsighted, the product people were tuned into processes that the engineers did not understand – and that was the stuff that was necessary to get things done.

More Changes

Just when the scientists were getting used to working with product developers, Lou Gerstner's IBM was changing its business model from hardware to services. Lou Gerstner, the consummate salesman wanted to sell everything that IBM had on its truck and he wanted things that nobody ever thought about before to be on the truck. It is not that hardware did not count for Gerstner. In fact, in the early 1990's it represented over 60% of IBM's revenue while services were just about 4%. But, IBM was losing market share in major hardware areas and Gerstner did not know how to stop the revenue drain. Gerstner needed to fix it. He actually needed to fix IBM by fixing it. Rather than change the whole company and risk having it go under, he decided to create a new business within IBM – a for-real services business.

Now, twelve years later, hardware is about 35% and Services is approaching 50% of IBM's revenue. Software has the other 15%. So, shortly after one cultural shift, IBM's research and development team had to endure another. They thought they were a hardware group but Gerstner saw everybody as part of a big services team. He even wanted IBM managers billing for their time. The willing engineers and scientists were quickly placed on the services team. A number of the talented human beings in IBM's Research Division and its Technology Divisions were then enlisted to provide research and development services for IBM's customers and competitors alike. In many ways it was one of the smartest things that IBM has ever done and it has been very successful for the company. Many companies have benefited from IBM's new ability to sell engineering services and these include AMD, Sony, and Microsoft.

Engineering as a Service

IBM now has a unique services offering that nobody in the consulting field can come close to. The IBM consulting team can offer something their competitors can't, the potential to assign an IBM scientist to work on their customer's problems. Right now, IBM's highly skilled business analysts and software scientists lend their expertise to such tasks as helping customers streamline their supply chains and analyze information on the Web, while the hardware scientists help clients perform exotic things like adding more parallelism or some IBM-only cooling tricks to help client chips provide better performance.

Additionally, when one of thc 1300 IBM scientists on the "away" team goes out on the road, they bring a pretty big suitcase full of IBM tricks. In the case is access to all of IBM's patents and other intellectual property. If you've got the "away" team on your site and you need something for which IBM holds the patent, you can get it in the bundled services deal.

That's not a trivial ability. IBM's strong research organization has complemented its crackerjack consulting organization. Even with IBM earning revenue from its engineers and scientists to satisfy the customer priorities, Big Blue still dedicates a substantial portion of those dollars (upwards of 15%) to the exploratory science for which the company is famous. Remember the ASDD and the SDD that IBM was so proud of through the 60's, the percentages may not be the same today but the notion is very much the same. The vast majority of researchers are three to five years out, but there's still room for a nice pair of sneakers and a tee shirt and a beard for those pie-in-the-sky projects that may never turn into products.

Without spending much more time in the topic of research, for impact and impact alone, I have provided below a list of the various subsets of materials research in which IBM is focused today. Materials are the key to unlocking the secrets necessary for the next generation of chips and the one after that. It helps to know that IBM is a world leader in technology and its engineers and scientists love making great things happen as much as they like getting on their Harleys at the end of a long day of thought.

Materials Research Is Key to the Future

Here are the types of things that IBM is working on in the Materials area:

Materials Science, Nanotechnology & Nanoscience

- ✓ Dielectric Materials
- ✓ Electrically Active Organic Materials
- ✓ Lithographic Materials
- ✓ Magnetic Materials/Magnetism
- ✓ Material Patterning
- ✓ Electron Microscopy
- ✓ Nanostructured Materials
- ✓ Superconducting Materials
- ✓ Theory and Computational Science
    - Algorithms and Theory
    - Atomic Wires and Molecular Devices
    - Carbon Nanotubes
    - Chemical Kinetics Simulation
    - Computational Science Collaborations
    - Correctly Rounding Math Library for Java
    - Dislocation Dynamics
    - iBoot – Remote Boot over iSCSI
    - Molecular Dynamics Simulations
    - Nanotube Theory
    - Quantum Cryptography
    - Quantum Information at IBM Almaden
    - Quantum Logic Experiments with Trapped Barium Ions

Licensing Innovation Pays Off Big Time

Just a few years ago, IBM was the first company to be granted over 3000 patents in one year. If as an organization, you do not feel that you need IBM's services but you do not want to re-invent the wheel, since IBM probably has a patent on the wheel you need, IBM still may be able to help. It doesn't matter whether you are usually a client or a competitor, in what IBM calls a spirit of coopetition, you can license the company's intellectual property, or as the company likes to call it, IP.

To put IBM's patent portfolio in perspective, over the last ten years IBM was granted in each year more US patents than any other company. During that time period IBM inventors amassed approximately 25,000 patents – more than the combined output of the company's ten largest competitors.

On the financial side IBM earned about $2 Billion last year from its IP portfolio. So, Big Blue not only does basic research from which it benefits in its products, it actually brings back about 40% of its research by licensing the fruits of its efforts.

Have Your Cake and Eat It Too

Licensing IP gives IBM the opportunity to sell something and still have it. Additionally, it gives the company the ability to profit from products that other people make. For example, IBM has been able to profit from inventions that the company is not prepared to market, such as excimer laser surgery to fix eyesight. That technology was developed by IBM in Yorktown Heights, but it isn't a business IBM is ready to pursue. At the Mayo Clinic, IBM scientists and Mayo Clinic doctors teamed to build a computer based sleeve that can work with an imaging unit so that the patient can sit outside instead of having to be wheeled into a small tube. Again, IBM will not market this device but its customer will and Big Blue again will benefit from selling its IP.

With all of its patents in the US and outside the US, IBM has a lot of intellectual property that other companies want. That means that the quiet chip company from East Fishkill NY and elsewhere that claims it is not at war with Intel, can forge lucrative cross-licensing deals with companies such as Intel or AMD or Microsoft, or Samsung, or

Nintendo or whomever it chooses. In order to play in the high technology game today, you must innovate and IBM's record places it well above the pack. Of course a major side benefit to all of its IP is that instead of having to pay real money for other companies' innovations, IBM can barter its intellectual property to the tune of many billions a year,

Model Research Program

There is no company like IBM today and industry observers suggest that IBM has become a model for corporations that want to reinvent their research programs. Are you listening Intel? Perhaps it was the fact that IBM had to sell whatever it could in the 1990's, especially things that could be sold and still be owned. Perhaps it was Lou Gerstner's phenomenal marketing mind. Either way, IBM taught the business world a lesson in the last ten years that there is more than one way to profit from your technology.

Learn By Mistakes

For a company that was literally in the toilet ten years or so ago, this is an unbelievable achievement and in so many ways it is Lou Gerstner's legacy. While it was approaching the lean years, like other corporate labs, IBM originally looked inward for its ideas. IBM thought it knew what was best for IBM. IBM thought it did not need to look elsewhere. It seemed to make sense at the time. Some might call that arrogance but others might just call it tradition. Hey, fifty years ago when IBM entered the computer business in order to do anything, the proper company credo was that you had to do everything. There was no choice. As times changed, IBM did not change, and in the 1990's it all caught up with the company.

The IBM of the 1980s and 1990's made a lot of mistakes by being insulated from what was happening in the industry. There was a time that IBM was the industry or at least IBM believed it was the industry. So, it is understandable but not necessarily forgivable that Big Blue did not capitalize on many of its own inventions. As noted previously, IBM invented RISC and Sun Microsystems implemented it and became the leader in this IBM-invented technology, and subsequently Scott McNealy became a billionaire. So also for Oracle. IBM's Tedd

Codd published all Oracle needed to know to beat IBM out the door with relational database technology. IBM was not interested in it because its big mainframes did not need it at the time and IBM did not want to interfere with sales of their other database types. For IBM's hard work, Larry Ellison became a billionaire.

From my own experience in the company, having lived through the 1980s and early 1990s with IBM, I see a new and smarter IBM. Though not as entrepreneurial as it can be, IBM no longer sits back and gets pushed around. The new IBM is crafty and cunning and in many ways is a reflection of its founder, Louis V. Gerstner Jr. Today's licensing and cross-licensing technology deals are a by-product of Gerstner's basic philosophies. They are also necessary for companies to tap each other's innovation and because of Gerstner and a cadre of brilliant scientists; IBM gets tapped more than it taps.

Let's not forget a by-product of the services game. IBM's engineering and scientific services pay the company back in more ways than revenue. The skills that IBM gains in doing things that have never been done before for its client accounts stay with the company and make it easier for the new IBM to get the next big score.

Open Innovation

Finally, IBM is in the forefront of something called "Open Innovation," in which the company, through its research division, effectively reaches out and connects to other centers of learning in both academic institutions and in other enterprises. Clearly the days when IBM's inward focus prevented it from taking advantage of its own inventions are over and Big Blue understands its leadership role in providing a formula for global laboratories and global fabrication plants in which the corporations and the institutions partner for their own betterment as well as the betterment of the whole. IBM's Power.org (www.power.org) and its Power Everywhere (Chapter 15) campaign are fine examples of how the notion of open cooperation and competition is paying huge dividends for the advancement of science and all of the participants.

Albany Nanotech

Intel submits that it accomplishes "pure research" through its affiliations with universities and it suggests that seed money is often a motivator to the researcher and / or the institution. Intel has also begun to look outside the US for its partners since the US universities are getting smarter about giving up intellectual property. If results are to be measured in products, and / or product improvements, from Intel's record of the last five years, however its research work with universities is not working.

IBM of course has arrangements with lots of different universities, just as Intel, but IBM does not rely solely on these universities for its research results. Remember, IBM has over 3000 researchers who receive IBM paychecks twice a month. Having postured that, IBM does not seem to be as hung up on who gets the IP as Intel as witnessed by the company's huge investment in the Albany Nanotech.

Albany NanoTech is a university-based leading edge global research, development, technology deployment and education resource facility. The Nanotech complex's mission is to support accelerated high technology commercialization. It is co-located with the new College of NanoScale Science and Engineering at the University at Albany (SUNY). Albany NanoTech is in technology big time. The organization seeks to leverage resources in partnership with business, government and academia to create jobs and economic growth for the nanoelectronics-related industries.

This research facility caters to small, medium and large industrial partners and provides them with access to state-of-the-art laboratories, supercomputer center, shared-user facilities and an array of other scientific centers. These facilities are intended to serve the partners long and shorter-term technology development needs while training their workforce for the 21st Century.

Albany Nanotech strives to provide a virtual "one-stop-shop" for its partners by assisting companies overcome technical, market and business development barriers through technology incubation, pilot prototyping and test-bed integration support leading to targeted deployment of nanotechnology-based products. Though training and tech-incubator space for start-ups is key, the organization also aims to pull together public and private funding to support R&D in microelectronics, nanotechnology, optoelectronics, bioelectronics and telecommunications.

In support of its mission, the University sponsored organization built what it calls NanoFab 300-South in September 2001 and took occupancy in January 2003. This facility is a 138,000 sq. ft. technology acceleration facility providing for business incubation, classrooms for the new College of NanoScale Science and Engineering, workforce training, offices for Albany NanoTech, and large and small industrial sponsors and partners including IBM, TEL, GE, and International Sematech North. The facility also includes 16,000 sq. ft. of clean room to support the International Sematehc North, IBM and other "next generation" nanotechnology research activities.

The pouring of the foundation for NanoFab 300-North began in September 2002 and the facility was ready for tool installations by the beginning of 2004 and it is now online. This new facility is a 228,000 sq. ft. It provides a Class 1 capable 300 mm wafer R&D, pilot prototype, incubation and workforce training and it also houses a 300 mm wafer nanoelectronics process line. The 350,000 plus sq. ft. complex also has over 50,000 sq. ft. of clean room space supporting the nanoelectronics-related industries.

IBM Is a Major Nanotech Sponsor

IBM is right there in the forefront of this revolution. The company initially invested $100 million in the Albany NanoTech center. IBM is also a member of International Sematech, a consortium that has put $160 million into a research project at the NanoFab and IBM is also one of the investors backing the Tokyo Electron Limited's research, a $200 million commitment to be matched with $100 million from the state.

Pure Research without the Cost

"We want to be two to four generations ahead of the industry," said Alain E. Kaloyeros, a physics professor at the University of Albany. Kaloyeros heads up an umbrella organization for the project, called the University at Albany Institute of Materials (UAIM). Albany NanoTech is intended to serve as "a Switzerland" of pre-competitive research for the industry, helping chip makers establish technologies and processes "before they go to customers with new products," Kaloyeros added.

The University at Albany launched the School of Nanosciences and Materials to complement the UAIM, which covers a number of programs, including the New York State Center for Advanced Thin Film Technology (established in 1993). UAIM also covers the Energy and Environmental Technology Applications Center, the Microsystems Integration Laboratory (for applications in nanosystems and micro-electromechanical systems, or MEMS), the Public Protection Technology Applications Center (sensors for law enforcement), the Center for Nanosciences Workforce Training, the advanced computer modeling program for nanosystems and processes, and the business outreach and technology deployment program.

In addition, Albany is one of the American universities involved in the National Interconnect Focus Center program managed by the Microelectronics Advanced Research Corp. (Marco), a wholly owned subsidiary of the Semiconductor Research Corp. (SRC).

"We want this to be a national center for workforce training," said UAIM's Kaloyeros. "New York State companies are a top priority for us, but even New York State companies need a national network."

Building on Work of Consortium

The University at Albany has been licensing equipment from IBM for the facility under a three-year agreement between the university and IBM. In return, IBM has access to the 300-mm wafer facility for its R&D projects. IBM is in the process of completing its own $2.5 billion project adding an Annex to its 300-mm fab in East Fishkill, N.Y. Already the Nanotech and IBM have produced some major inventions in immersed photolithography which have the potential to ass additional capabilities to this critical area of chip making.

IBM has internship slots for 20 Albany students to study at various IBM facilities. And, the company will support Albany's research activities through grants and equipment donations in its University Partnership and Shared University Research programs.

In addition to licensing production equipment to Albany NanoTech, IBM provides "guidance" to the university center on other equipment purchases, Kaloyeros said.

IBM recruited 50 graduates from this year's senior class at the University at Albany which creates for IBM a pipeline of students that will increase over time. Though Albany Nanotech is not just targeting IBM, the organization is quite happy with the help it is receiving. And IBM is quite happy with Albany Nanotech.

Governor Pataki Gets Lots of Credit.

There is no university in the world with the capabilities of Albany Nanotech. To a large degree, Mr. George Pataki, NY State Governor has staked his legacy on what is happening in the buildings there. Pataki's dream is to turn the old Erie Canal corridor into a place where new technologies are developed and futuristic products made. Pataki is not only looking at this center and others in NY State to make scientific breakthroughs, but he is also looking for the Nanotech to build prototypes.

Pataki put his money where his mouth is. While New York's Senators have been studying the loss of jobs in NY State, Pataki's administration itself has been doing. The state itself pledged $620 million for the next five years to help develop technology that could be turned into commercially viable products. In the past, most of the cash for research into chip-making and other scientific advances came from the Federal Government or the private sector. Not surprisingly, executives in the semiconductor industry applaud the state's contributions. In fact, Mr. Pataki was named "Man of the Year" in 2003 by the Semiconductor Industry Association in San Jose, Calif., in the first presentation of the award to an elected official.

"He's way ahead of the curve relative to the other states," said an I.B.M. senior vice president, John E. Kelly III.

Lithography Research

The long-term dream of Albany NanoTech is to create a fully integrated resource for a variety of disciplines, a veritable one-stop shopping tech mall for R&D, education and technology deployment. The funding is well on the way.

On July 19, 2005 for example, at its kcickoff, IBM and AMD started pumping even more money into a more than $600 million public-private semiconductor research consortium to be established at the University at Albany in New York. Also "chipping in" money for the project are Infineon Technologies AG, Micron Technology Inc. (NYSE: MU) and of course, the state of New York. The project is known as the International Venture for Nanolithography, or INVENT.

AMD, with a similar slant on partnering as IBM was represented by Craig Sander, corporate vice president of technology development who stated: "Strategic partnering is one of the keystones of AMD's overall technology R&D program and a critical element of what allows AMD to be a leader in semiconductor technology. The INVENT alliance is an ideal example of partnering for mutual advantage."

John Kelly, senior vice president of technology and intellectual property at IBM, offered: "IBM congratulates New York State for its support of this landmark collaboration of global leaders in semiconductor development and manufacturing. This joint investment further establishes Albany NanoTech as a center of excellence in advanced lithography research, and an international magnet for investment, jobs and innovation."

Honeywell International Inc. is also part of the game as it recently announced plans to set up a research center at the Albany University site. Tokyo Electron Ltd., a manufacturer of tools used to make computer chips, and International Sematech, an Austin-based consortium of computer chip manufacturers, already operate research centers there.

Dr. Wilhelm Beinvogl, Senior Vice-President for Technology Innovation at the Memory Products business group at Infineon Technologies said, "Understanding the challenges of the next generation of advanced lithography and developing the right tools is very important to the semiconductor industry. This partnership is a key effort to develop that expertise and keep Moore's Law on track."

"The INVENT program is an example of the power of partnership when endeavoring to support pre-competitive research that will have a favorable impact on society's productivity. Micron Technology is committed to participating in the effort in a way that leverages the partnership," said Mark Durcan, Micron Technology, Inc.'s CTO and VP of R & D.

My guess is that over time, there will be more Universities with Nanotech-like partnerships to industry. But, this one has taken a good five years to get where it is, and it is no coincidence that IBM's corporate headquarters as well as its major chip making facilities are in New York State.

The world's largest chipmaker, Intel, however, is not on the short list at Albany Nanotech, but the world's technological leader in x86 architecture, AMD is one of the lead participants. Maybe, just maybe, that's part of the Intel R&D problem.

Chapter 25

The Real War - IBM v. Intel

The Hustler

Without declaring a war on Intel, IBM has recently been out-hustling the most dominant chip maker in the industry. Though typically quite demure in its marketing and its competitive strategies, chip folks at IBM know they have all the chips covered with a very strong Power Architecture, a very game Cell processor, the best ASIC chips, and a deep partnership with AMD. All of this adds up to IBM being postured to out-Intel Intel in all areas, including the chip giant's bread and butter area – high speed x86 processing.

Intel Outside

From the time of the IDF in 2002, Intel had been struggling to ramp up the clock speed on its chips and until late 2004, it apparently believed that it could. It was always a tried and true method of jacking up performance without having to study all of the potential for on-chip improvements. The approach had never failed until the Pentium 4 met its challenge with heat.

The notion of continued high clock speeds in Intel mainstream chips died a slow death over a three year period in which the Santa Clara chip maker could not raise the performance bar but a hair. In late 2004, the notion of high clock speeds as the best way to enforce Moore's Law was taken off the books inside Intel. When the clock stopped, Intel was on the outside looking in. The company had not invested in other ways to make its chips purr, and many companies lost respect for Intel as an innovator. .

Overall, Intel lost three years and perhaps more, in trying to get more clock speed from its single thread chip designs. In the fall 2004 the company just gave up. The Intel Company notified the world that there would never be a 4GHz Pentium 4 chip. Instead, the chip giant was preparing a full 180 change in its chip design by adopting the notion of parallelism in its chips, a place where IBM's Power processors had been sitting very comfortably for over ten years.

IBM's POWER chip line is far ahead of Intel in the parallel processing competition. IBM started its work with chip parallelism in the early 1990's and the company has been doing multiprocessing for well over forty years. IBM is where Intel wants to go. IBM is the one and only company out there with the know-how.

Back Peddling to Parallelism

The fastest Pentium 4 chip of all time clocked in at 3.8 GHz. Due to an unfortunate heat spell, the performance corollary to Moore's Law stopped working for Intel. Being forced with the reality that it could no longer crank up the clock to get performance boosts, Intel was forced to back peddle into the idea of doing more than one thing at a time on a chip. So, now parallelism is a necessary means for Intel to bring more power to its chips. Intel has no choice.

Clock speed has never played an important role at IBM. First of all RISC clock speeds are typically less than CISC though and RISC systems get more done in one cycle than CISC designs. Rather than clock frequency, IBM relies on industry standard performance benchmarks to gauge the power of its systems. Benchmarks determine how much work a chip can do rather than how fast a certain aspect of the chip, such as clock speed may be. IBM's POWER chip clock speeds have not been tremendously high, but because of a substantially more innovative chip design than Intel's approach, the work accomplished per machine cycle in POWER chips has always been dramatic.

Other Factors

To demonstrate the dramatic significance of factors other than clock speed, consider the clock speed difference between the IBM POWER4 and POWER5 processors at announcement time. The difference was a mere .35 (35/100) GHz. Intel chips could spit out .35 GHz and not notice the difference. Yet benchmarks between POWER4 and POWER5 showed that the POWER5 chip ran better than double the performance of the POWER4. Some benchmarks place POWER5 at four times that of the POWER4. How can that be?

Clock speed was the easy way to gain performance and if it had continued to work, many analysts think Intel would not be embracing parallelism right now. In each processor family iteration, IBM always has done more than merely ramp up the clock speed. Silicon on insulator (SOI), copper interconnects, dielectric "k," more threading and more intelligent threading, and whatever the lab can get on the chip to make it better– that's what IBM does with each planned generation of Power Architecture.

IBM also has a trick going on in its chips called "Fast Path." Power5 Fast Path for example takes over software tasks commonly used in the operating system such as packaging data to be sent to networks. When POWER6 comes along, it will extend its reach further, taking over tasks now handled by higher-level software such as IBM or Oracle database software or IBM's WebSphere e-commerce software. Things that once needed operating assistance will now execute in a few instruction cycles.

Benchmarks prove that you can get lots of performance benefit merely by designing the next chip better. But, if the world evaluates your wares on listed clock speed alone, and Intel was always the clock speed king, on specs alone, your chips could not appear to be as fast. Again, those days are gone.

Parallel processing with its bread and butter x86 chips is an area in which Intel was not seriously engaged. It was an aspect of performance that the Santa Clara company was implementing with HP for Itanium but the chip giant had not felt it necessary to explore parallelism in any big way in its x86 line. As we have reported consistently in this book, so far, Itanium so far is taking nobody anywhere.

From the eyes of many industry analysts, including yours truly, the IBM POWER5 is the best chip in the industry and IBM intends to keep the lead. Look no further than to the IBM mainframe division's decision to move to the Power Architecture. IBM's mainframe division would never consider moving to POWER if it was not assured that it was the highest performance technology for the future.

Apple is the one anomaly. The Apple switch to Intel may not be good PR for IBM, but it is much to do about nothing. On the other hand, if the very independent mainframe division of IBM had chosen Intel over POWER, now that would be reason for concern.

POWER Produces "Cool Chips"

The POWER5 chip is at the heart of the new IBM i5 (iSeries) and its cousin the p5 (pSeries). The POWER6 will be at the heart of these as well as IBM's next generation mainframe when it is introduced in a year or so and maybe Microsoft XP will also be there as a choice. IBM uses some nifty technology tricks, many of which were learned because the company also builds mainframes, to gain the performance advances and handle the cooling requirement.

According to Dr. Frank Soltis, IBM has in fact solved the heat dissipation issue. Unlike Intel, IBM does not try to heat up its chips to get more power and then hope the cooling guys can keep up with the heat. Instead, IBM designs its chips with lots of tricks to keep them cool, and it calls in the cooling guys only if the tricks don't work.

One of the most publicized tricks in IBM's arsenal is to make the chip itself smart enough to turn off parts of the chip that are not being used. How does IBM do this? In addition to hardware, IBM puts some smart software on the chip that looks far ahead in the instruction stream to recognize areas of the chip that are not about to be used. After knowing what is not needed, this IBM smart technology shuts off power to the areas of the chip that will not be needed several instruction cycles down the road.

This has the affect of conserving lots of electrical power but more importantly, because chip circuits are not being lit up unnecessarily, the chip runs cooler. Consequently, IBM can run its chips faster since there is less concern for the chip overheating and producing erroneous results. IBM has shared the smart shut-off technique with the public, but the company does not even talk some of the secrets it uses to keep its chips "cool." Why would IBM want to teach competitors who have never cared about heat, how to deal with heat?

Is IBM Ready to Take on Intel?

To a neophyte studying supply chain management (SCM), one of the obvious sticky areas is when your supplier is also a competitor. The Power Architecture is definitely in competition with Intel for market share and mind share in the chip processor space. In recent years, from the eyes of most industry analysts, IBM has gingerly competed with Intel. IBM has avoided any head-on competition for the heart of Intel's business, the x86 PC processors.

IBM is not just a chip manufacturer. IBM is a system, software, and service supplier. As a system supplier, one of the systems that IBM sold was the PC. As a PC manufacturer, IBM depended on Intel for its supply of Pentium chips. It would be rather difficult to be too independent of your supplier when you are in need of supply.

ASIC chips and Game Controllers

Of course IBM has been OK for years in other areas of the chip industry, such as those areas in which the cost of Intel chips often precluded their use, independently of IBM competition. For example, IBM has been the ASIC (embedded processors) technology leader for some time. Intel chips were never cheap and IBM priced its chips at the market line so it could get the business. The great home entertainment vendors could not make money using Intel's expensive PC chips so they looked to IBM to satisfy that need. So, today and tomorrow and in the future, you will see IBM in video game controls, embedded processors in microwaves, automobiles, cell phones, etc. IBM has been aggressively pursuing many sales opportunities for its POWER line while avoiding a head-on clash with the fab foundry leader, the most prolific chip manufacturer in the universe, Intel. But, how long can that last? How long will it be until, in a defensive move, Intel declares war on IBM?

IBM Played Softball with Intel

IBM had good reason to wear the kid gloves. Having a competitor (Intel) as a supplier of chips for its PC line made IBM a bit nervous while the company continued to push its PC consumer wares. After all, if there were any question as to whether a loyal Intel customer such as Dell or a chip competitor such as IBM would get delivery -- if supply ever got tight-- IBM well knew that you could place your money on Dell. In an active and necessary supply chain, that is not a comforting thought.

So, IBM's PowerPC and POWER initiatives that may have taken on Intel in neutral surroundings were quite muted over the years because of the IBM PC Division's dependence on Intel chips. The "Intel Inside" mantra had a big impact on how Big Blue's PC Division behaved.

AMD Inside?

Of course, within IBM and outside, many, including yours truly have been urging IBM to get out of the PC business for years because it is a distraction to its strengths. IBM has been examining all of its options regarding Intel for some time. In 2003 when AMD and IBM teamed up, it marked a subtle change for IBM in its Intel relationship. Intel's major competitor was AMD and IBM was helping AMD. However, because Intel always expected to use Moore's Law to win the game in the end, even that was OK. But, when AMD was actually able to out-engineer Intel and out-market Intel, the game got a little more difficult for IBM. Big Blue had to make a decision – were consumer PCs or industry leading processor chips more important to IBM's future success? IBM chose the latter.

The three press snippets below give an appreciation of the level of cooperation that exists between AMD and IBM to this day and it is expected to last beyond.

Snippet 1: http://www.newsfactor.com/perl/story/20420.html

AMD Teams with IBM on Next-Gen Chips

By Keith Regan
NewsFactor Network
January 9, 2003 10:52AM

"Aberdeen Group analyst Peter Kastner said AMD may be searching for a competitive toehold against Intel, which recently has begun to erode AMD's gains in the low-end processor market.

AMD has inked a pact with IBM to jointly develop next-generation microchip technologies, with an eye toward making faster chips that use less power.

The two companies said they will focus on advanced chip structures and materials, such as silicon-on-insulator (SOI) transistors, copper interconnects and improved "low-k dielectric" insulation. They plan to produce 65- and 45-nanometer chips, which will be made available on 300-millimeter wafers. Much of the work will focus on improving the processes by which the chips are made.

Right To Work

Work is scheduled to begin by the end of January, with initial development taking place at IBM's Semiconductor Research and Development Center in East Fishkill, New York. However, the companies said their first 65-nanometer chips likely will not become commercially available until at least 2005."

Snippet 2 http://www.linuxinsider.com/story/36834.html

AMD and IBM Extend Chip Processing Deal

By Jay Lyman
www.TechNewsWorld.com
Part of the ECT News Network
09/22/04 11:22 AM PT

"The chip collaboration highlights the barriers to advancing chip designs, allowing both companies to build on their existing investment in manufacturing processes and keep up with Intel, which is becoming the only company capable of going it alone.

...

AMD said in its SEC filing that it had, through a letter of agreement with IBM, signed on to continue jointly developing logic process technologies, which include 65-nanometer and 90-nanometer chip manufacturing.

AMD said that during the three-year extension it would pay IBM between $250 million and $280 million, to be dependent on the number of partners engaged in related development projects at IBM facilities.

The parties additionally agreed to extend the target dates for achievement of certain development milestones, the SEC filing said."

Snippet 3
http://www-03.ibm.com/chips/news/2004/1213_amd.html

AMD, IBM announce semiconductor manufacturing technology breakthrough -- Industry-first advancement can improve performance, conserve power in future single- and multi-core processors.

Sunnyvale, Calif. and East Fishkill, NY, December 13, 2004
AMD and IBM today announced that they have perfected a new strained silicon transistor technology aimed at improving processor performance and power efficiency. The breakthrough process results in an approximate 24 percent transistor speed increase, at the same power levels, compared to similar transistors produced without the technology.

Faster, more power-efficient transistors are the building blocks of higher performance, lower power processors. As transistors get smaller, they operate faster, but also risk operating at higher power and heat levels due to electrical leakage or inefficient switching. AMD and IBM's jointly developed strained silicon helps overcome these challenges. In addition, this process makes AMD and IBM the first companies to introduce strained silicon that works with silicon-on-insulator (SOI) technology, which provides additional performance and power savings benefits.

"Innovative process technologies such as strained silicon enable AMD to deliver more value to our customers," said Dirk Meyer, executive vice president, Computation Products Group, AMD. "Our shared progress in developing advanced silicon technologies allows AMD to deliver today's best performance per watt, and this strained silicon development will extend that leadership when we begin shipping the dual-core AMD Opteron processor in mid-2005."

Future of Computing Defined by the Outcome

No matter what side of the fence you are on in the AMD v. Intel or the potential IBM v. Intel battles, you've got to admit that the future of computing rests within the outcome. Since IBM now appears ready, though it denies it publicly, to take-on Intel in the chip processor battleground, along with other business reasons, IBM made the decision to drop its PC business. It chose the chip business. So, now other than for some xSeries servers, which do well being powered by AMD, IBM does not have a major need for Intel on the supply side.

IBM viewed the marketplace reality of the mid 2000s and acknowledged the success of "Intel Inside" in the home market. For IBM to stay in the PC marketplace, it would have to continue to treat its primary x86 chip supplier in friendly terms. Personal PC users and even light corporate PC users were not interested in PCs that were not emblazoned with "Intel Inside," even if it might be OK with IBM.

"Intel Inside" has not been enough to sway the pros in the server ranks, however. Intel's ingenious slogan simply does not play as well with IT professionals. Many are under-whelmed by marketing blurbs such as "Intel Inside." Whereas a neophyte may have a hard time understanding that AMD inside might be a better deal, a server professional typically understands the underlying technology and the performance factors. With AMD's Opteron chips out-performing Intel in traditional server benchmarks, and with the ongoing IBM and

AMD partnership, the server professional has no reason to hold any "chips": against IBM or AMD. It has been easy to see for almost three years that the AMD Opteron Chip, made with some level of consortium with IBM is the unabashed leader in Xeon class x86 server performance.

AMD in IBM Servers, Not PCs

Just a year ago, IBM began to adopt AMD chips in its sever line of x86 units but for its workstation and Laptop PCs, Intel was its sole supplier. From looking at the last ten years, it is clear that IBM's PC Division was not really sure whether it could embrace AMD or whether it had to give all its love to Intel. Periodically over those years IBM would start a relationship with AMD for PC sales but overall, since IBM closed AMD PC deals as quickly as it started them, IBM's view was that overall; AMD would hurt its Intel supply chain. When Intel was not pleased, IBM's PC Division was very uneasy.

To keep the supply chain running well for its consumer PC line, IBM would abruptly stop making AMD–processor based PCs or end deals that were in process. If AMD's June 2005 lawsuit is to be believed, Intel had monopoly power over IBM to assure that even the "mighty IBM," could not stray too far from the Intel fold – even if it wanted to stray. That's what you call supply chain power or monopoly power depending on your point of view. With IBM as the customer, without getting the government to intervene and IBM was not about to do that, the company was powerless if it wanted to sell Intel products.

No AMD Units in IBM's Consumer Line

To verify that my research on IBM's use of Intel and AMD was correct, in early April, 2005, I used my IBM retiree status to check the IBM PC wares that the company continued to sell to its employees. There were no AMD processors in the consumer PC model mix. Not one! IBM had shied away from AMD in the consumer PC marketplace to preserve its supply chain interests by keeping Intel happy.

In his COMMON presentation on the future of IBM technology, Dr. Franks Soltis noted that IBM had chosen for years not to aggressively compete against Intel at the chip level since the company clearly was dependent on Intel for a supply of Pentium and Celeron chips to make its PCs. Now that IBM has sold its PC Division to Lenovo, Big Blue no longer has such supply chain constraints. Moreover, the company now behaves in ways that demonstrate that its chip business is very formidable—formidable enough perhaps to take on Intel after all these years.

What Chip Actions Can We Expect?

The IBM company has clearly decided that it can do better today with processor chips than with consumer PCs and so it was a natural act for IBM to decide to sell its PC business. Now, as we move to the second half of the first decade of the new millennium, since Intel is no longer an integral part of the IBM supply chain what can we expect? As you will see, IBM is competing with Intel for everything, not just server chips. We can expect more of the same.

IBM Is Tops in ASICs

Quite simply without any public announcements, fanfare, or rhetoric, IBM is aggressively pursuing Intel's chip business by marketing to the same customers as Intel. With one industry leading scientific advancement after another, the IBM Company has not been sitting still over the last few years with regard to chip technology. Instead IBM has been going after chip business in all areas in a big way.

For example, IBM has developed a major market for itself in selling inexpensive chips to companies that need powerful technology for use in their own products. IBM is very big in application specific integrated circuits (ASIC) and it is about to be the unquestioned leader in game consoles.

We have not examined the notion of ASIC in any great detail in this book. In essence these are part standard and part customized processor chips. The ASIC chips are designed by chip customers, not the manufacturer, for specific applications by integrating technology from standard sources of pre-tested code. ASIC design is faster than designing a chip from scratch, and design changes can be made more easily. ASIC chips are the heart of the technology used in embedded systems in wireless phones and infrastructure as well as direct broadcast satellite (DBS) systems, very small aperture terminals (VSATs), radar systems, digital and mobile radios, synthesizers, voice storage systems, security systems and instrumentation.

ASICS can be a very big part of a chip manufacturer's business. Intel's CEO Barrett in 2005 announced that Intel was heading for big success in the Cellular industry which depends on ASIC chips big-time. Intel would love to capture this market but it has eluded the chip giant so far. So, in another area of competition, the ASIC chips for cell phones, IBM is winning market share as the unquestioned technology leader in the ASIC industry.

Demonstrating its leadership in ASIC and getting a big jump on the leading edge chip market, on June 13, 2005, IBM rolled out a pair of 65-nm application-specific integrated circuits (ASICs), including a low-power offering for the wireless, mobile and consumer electronics markets. The new chip, of course includes the IBM advanced features of copper interconnects, low-k dielectrics, strained silicon and other advanced chip features.

IBM announced that this new chip technology reduces leakage by up to 30 times from the company's previous 90-nm ASIC offering and it increases performance by as much as 20%. Tom Reeve's IBM's OEM Vice President, Systems and Technology Group praised the new chip in his statement: "This dual offering will continue our undisputed logic ASIC technology leadership at the high end. Now with the introduction of our low-power ASIC offering we will bring our proven record of technology leadership to the high growth consumer marketplace."

Microsoft Loves IBM Again

As difficult as it may be to believe, even Microsoft Corporation has joined the POWER camp and will be building its new Xbox 360 machine using multiple IBM POWER Processors. Gutsy IBM went out and stole one of Intel's big customers. You can bet that if IBM was not the best in the chip business, Microsoft would have stayed with Intel. Unlike Apple PCs, the Xbox 360 as a relatively inexpensive consumer product is expected to ship in high volume.

The gloves are certainly off in the chip wars. It is fairly certain that if IBM were still in the PC business and it needed Pentium or Celeron chips to round out its PC line, stealing one of Intel's big customers (Microsoft) would not be the way to achieve supply. The following snippet from Computer Reseller News at vnunet gives a great perspective on IBM's latest POWER conquest – the unconquerable Microsoft.

http://www.crn.vnunet.com/news/1151476
IBM PowerPC chip gets gaming boost
Big Blue's 64-bit chip set wins consumer and enterprise deals
Peter Williams, **vnunet.com 11 Dec 2003**

"IBM is picking up deals for its PowerPC 64-bit chip set, intensifying its rivalry with Intel's Itanium. Last week Microsoft chose the PowerPC chip for its Xbox 2. The software giant is also using the the 64-bit chip extensively in its own enterprise servers alongside Intel processors, and possibly AMD in future, for 32-bit applications.

The Microsoft agreement follows original equipment manufacturer deals for the Apple Mac, Nintendo Gamecube and Sony PlayStation.

David Valentine, IBM's Linux EMEA sales manager, said: 'The PowerPC is designed for very high floating point applications and is suited to rendering for animation, making it ideal for gaming. The agreements give us economies of scale.

The same features are suited to specific enterprise applications.

'Scientific and technical customers requiring high memory bandwidth and low latency discovered they were reaching the performance edges of 64-bit Intel [Itanium 2]," claimed Valentine.

'The problem was overcome by using PowerPC-based clusters'"

Specifically, the next-generation Xbox 360 will be equipped with three PowerPC chips, each based on IBM's POWER5 or POWER6 architecture and fabbed at 65nm. That POWER connection means each will offer simultaneous multi-threading technology, allowing them to process two program instruction streams at the same time. With three CPUs in the box, two cores per die, that means the console has the equivalent of 12 processors inside – and that's a lot of processing horsepower.

IBM Cell Processor, Positioned to Win

The Sony, Toshiba, IBM Cell processor is staged to conquer the gaming world and beyond. It is already 10 times faster than current technology and the industry has taken notice.

With Cell and IBM's other POWER processors and its leadership in custom ASIC chips, IBM is poised to lead the game console industry and the application-specific industry. Moreover, as the game console becomes indistinguishable from the home PC, IBM is ready to visit Santa Clara to pick up all the marbles.

As noted in Chapter 15, there is nothing in the Intel product line, the Intel pipeline, the Intel design lab, or the Intel dream lab that compares with the Cell processor. It's one of a kind, and it's from IBM.

Rochester Minnesota: IBM's Cell Processor and Supercomputing Lab

IBM is bar none, the undisputed leader in supercomputing with BlueGene/L. IBM's leadership has already paid off in other areas such as the interconnects of the POWER5 chips as well as the on-chip interconnections of the nine processors on the Cell chip. As the supercomputing architecture evolves from many physical hardware systems with fast external interconnects to many single-processor chips with super fast internal interconnects to one multi processor super chip with extremely fast on-chip interconnects, it helps as the technology reaches the chip level for a company to have been part of that evolution. IBM technology now exists in all levels of the supercomputing hierarchy.

While Intel was making its living on stretching its PC power by increasing its single thread speed, as you can see with these supercomputing efforts, IBM has figured out how to manage millions of computers at once in parallel. And, if one day all million or more of those little guys fit on a chip or two, IBM's supercomputer research will have paid off in practical uses for its leading edge processor line

Right now, IBM owns the first commercial supercomputer on a chip (Cell) and IBM is the best in supercomputing according to Top500.com. As further circuit densities are achieved according to Moore's Law, one can expect that the company that could do it the best with multiple systems and multiple chips will also be the company to do the best job as supercomputing continues its movement to the single chip level.

Future of Chips

To put this chip era in perspective let's assume for a moment that it is 1915 and we are in Indiana. It's about time that we hear those infamous words first uttered in 1915:

"Gentlemen, Start Your Engines."

But, it looks like they are already started!

What about Intel?

Intel has fifteen new processor chip designs all set for next year. These will not try to rely on clock speed but instead will be dual core and Hyper-Threaded and they will feature 32-bit and 64-bit capability (EM64T). So far, Intel has not announced any other noteworthy advances that will jettison its chip performance to the stratosphere. But, Intel is a winning company so I would keep watching them. Now that they have had their AMD come-uppance, I would expect the company to remain agile for at least for the next several years. With all of its R & D, and with the shackles of Itanium loosened, I've got a feeling Intel will be announcing something soon and it ought to be good.

What about AMD?

AMD is already ahead of Intel in terms of chip power and recent innovation. With its arrangement with IBM, it is even possible that AMD may begin to use a RISC based POWER chip as the engine for its x86 architecture. Perhaps IBM can bring back the PowerPC 615 instruction set (Chapter 10). Since Intel and AMD both run the x86 instruction set on top of a RISC processor, it is not too far fetched that AMD would choose the best chip in the industry as the basis for its own new chips. All IBM would have to do is leave a little extra blank space on the chip for AMD to do its thing.

Moreover, with the IBM technology agreement, since IBM knows heat, you can bet that AMD will soon know heat and will begin to incorporate specific IBM-developed notions on its chips to keep them more cool and help them get more done in parallel. AMD is not in a bad position, having the leading vendor in chip technology as its partner. Additionally, AMD has its share of those 1300 IBM roving scientists who bring IBM's patent portfolio along with their services. Intel has big reasons for being concerned with what is being cooked-up in AMD's labs and/or in IBM at Fishkill N.Y.

What about IBM?

So what is next for the Power Architecture? Surely the Cell processor will expand whatever opportunities may be out there. Perhaps IBM will convince Sun to put Solaris on POWER and abandon its UltraSparc chips. Perhaps Microsoft will decide to put Windows on POWER. Considering that Microsoft has signed for the Xbox2, the company must plan to run a Microsoft OS on POWER to "power" the Xbox 360. What is the best OS that Microsoft has on its shelves for the future Xbox 360?

XP on Xbox2?

How about a Windows NT derivative such as XP? It would be very doable since Microsoft once had its NT OS running on Power PC in the 1990's -- before the company canceled the project. At that time, after NT had been made to run on PowerPC, Microsoft chose to run NT only on the x86 and DEC Alpha processors. So, it is highly conceivable that XP will be the driving OS on the Xbox 360. With IBM denying these rumors all the time, it seems even more likely that this will happen.

Sony Cell PC Workstation

Perhaps in the future Sony will use the Cell chip to introduce something more than a game console. Rumors are that Sony is planning to introduce a business workstation that is not designed for games. As a POWER unit, theoretically, the SONY would be able to run many operating systems such as OS/400, i5/OS, Unix, Linux, Windows XP, and even the mainframe operating systems as they are ported to POWER technology. That would not be a bad little workstation. If IBM wants it, another agreement with Sony could give Big Blue an inexpensive small-server platform running any IBM operating system, geared to meet the demands of the masses.

How Did IBM Get So Good?

The whole notion of Cell computing has given IBM a leg up in the supercomputing race as well as the processor chip race. Cell was actually a natural because to get real good at supercomputing, IBM had to learn how to build the high speed connections for computer chips to talk to computer chips, from intelligent Cell or POWER processors to other intelligent chips as well as the onboard chip cores of the Cell processor. In other words, by understanding supercomputing and developing technology for computers to talk to computers, IBM is poised to continue to make the components of supercomputing smaller and smaller and eventually, perhaps even a huge supercomputer will fit on a chip. Then, as one would expect, IBM would want to combine the chips into super-supercomputers.

IBM got good first because of its initial America project to build the RISC System/6000. IBM set out to build the best RISC chip of all time. Big Blue accomplished its America project objectives. Then, IBM kept getting better because of a number of concurrent and ongoing talents including the influence of Dr. Frank Soltis, the IBM corporate mainframe technology team, Bernie Myerson's team, past chairman Lou Gerstner's desire to be the best in supercomputing even if the objective were trivial (e.g. beating Kasparov in chess), etc. There are lots of other reasons why IBM has gotten so good but perhaps the most important is that IBM's scientific culture won the day in the corporation and thus IBM's executives permitted the excellence of their chip designs to be implemented in real products.

Now that we have a perspective on IBM's prowess in the chip industry, there is another big secret in IBM's backclosets. IBM has the most advanced operating system in the industry in what was once called OS/400 and now i5/OS. This OS runs on a POWER5 based i5 that in a book released in spring 2005, I dubbed the All-Everything Machine. If IBM one day chooses to mainstream this OS by adding a GUI front end and natural Web development facilities, then both parts of Wintel will have to look over their shoulder to see what is going on at IBM..

On top of all IBM's past accomplishments, Big Blue is far from finished. You can bet the innovation won't stop. With AMD as IBM's partner in the x86 race, with Cell and POWER about to own the game console space, and Power technology opening up to more and more manufacturers (www.power.org), and the "POWER Everywhere" initiative described in Chapter 15, in time Intel may feel very alone. All the work done in supercomputing and super chips is related and it is work done in IBM labs, not at Intel.

What's Next?

Can we expect that there will be an extension from on-chip interfaces and major league supercomputing clusters to other computers not so close by? With IBM knowing how to switch from processor to processor and chip to chip and local system to local system, these elements can be positioned in the near future to go from chip to local system or from chip to a remote chip someplace in the distance to get more computing power for today's job. Thus again, IBM's current developments in supercomputing and Cell computing on a chip, position the company well for when and if the notion of grid computing hits the marketplace in a big way.

In many ways, IBM is already the leader in grid computing for its current research and its "On Demand" initiatives have positioned Big Blue to take this next logical step.

IBM's Power Architecture Is No Reason to Go Anyplace Else

You may recall one of the unforgettable lines in Wendy's commercials over the years: "Wendy's ain't no reason to go anyplace else." Well, if you've gotten a look at the Power Architecture and you've gotten a look at where IBM and now all of its partners in power.org are taking the Power Architecture, you too would conclude that "POWER ain't no reason to go anyplace else."

When originally announced, the initial dual core POWER5 chips ran faster than Intel chips but at a modest 1.5 GHz and 1.65GHz clock speed. That's because IBM knows heat and performance. They are built using a 130 nanometer process. Before the end of 2005, IBM brought out its POWER5+ version with updated technology that uses a 90 nanometer process and offered speeds up to 3GHz. In 2006, IBM will move on to POWER6, which will use a 65 nanometer process (that is 65 billionths of a meter or not too many atoms) and the chips are expected to run up to 5 GHz. IBM is already into 65 nm chip building in its ASIC fabs.

IBM has its POWER6 chips in test mode in the lab right now preparing them for production next year. After POWER6, IBM will continue to release new versions every eighteen months or so. IBM's current roadmap goes out to POWER9, which should arrive in about six years or so.

IBM is very serious about staying in the chip business big-time and for the long haul. Its internal road map for POWER is laid out all the way to POWER9. By demonstrating consistency in delivering its chips on time, according to plan, IBM has surely struck some major fear in the hearts of its competitors, especially Intel. Just look at the ease with which IBM has implemented its new chip technology such as the first dual-core POWER4 chip way back in 2001. No doubt IBM is on a roll and, as you can see in the table below, the roll takes POWER way out into the future:

Continued Evolution of the POWER line:

- ✓ 2005: POWER5+, 90 nm Cu/SOI process
- ✓ 2006: POWER6, 65 nm Cu/SOI process
- ✓ 2007: POWER6+
- ✓ 2008: POWER7
- ✓ 2009: POWER7+
- ✓ 2010: POWERn.....

As a system manufacturer, IBM does not just stop at the chip in its innovations. For example, the POWER5 systems already scale from 1 to 64 processors. IBM expects gains over the short haul on its chips by improving floating point performance and simultaneous multithreading (SMT) technology. IBM has announced that it can turn the SMT technology off and on as needed, depending on the software load. And, with IBM supercomputers running with 1 million processors, why would we not expect to see 128 and 256 processor systems or larger within the next few years?

With POWER6, IBM plans "very large frequency enhancements," which again is not a good sign if you are IBM's competitors. The chip is due out in 2006 or 2007 and will be used in all non-Intel servers. By then, IBM is expected to use the chip in its mainframe line so all three server lines, the eServer zSeries, iSeries and pSeries will all run on the same chip. Those in the know say that the POWER6 chip is being developed under the code name ECLipz. It's not like IBM but this does sound like it is some kind of humorous Sun Microsystems reference.

With the probability that Microsoft Windows XP running on the POWER platform that powers its Xbox 360, it is highly likely that one eServer iSeries will be able to run Mainframe, Unix, iSeries, and Windows operating systems on one chip in micropartitions at the same time. That's almost hard to believe unless you've been watching what's happening with POWER.

IBM is well covered on the x86 front with its alliances, the POWER front, the PowerPC front, Cell front, ASIC front, and the Supercomputer front. I make no predictions because I am not a stock analyst, but the price of IBM stock may be the lowest it will ever be again if IBM marketing can figure out how to leverage all of its chips and secrets.

AMD's Future Options

AMD plays extremely well in the x86 space and if the second largest x86 chipmaker can overcome Intel's marketing machine through its own cleverness or through some help from the courts, with its partnership with IBM, the company has a great chance of overtaking Intel in the future x86 processor space. Of course with gimmicks like "Intel Inside" and "Centrino Inside" Intel is looking for lots more than just the chip business. It does not look like AMD is going to get beateb by Intel's research until the Santa Clara chipmaker reorganizes it to assure that it provides benefit. However, Intel has lots of marketing resources and AMD can be easy prey at the marketing level if Intel begins to spend.

An IBM/AMD merger would be an interesting event. A Microsoft/AMD merger would be a more interesting event. An Intel/AMD merger would be a real smart move for Intel. Considering the load of AMD debt, the company might go cheap. Out of all the possibilities, Intel stands the most to lose if AMD merges with somebody unless Intel is the partner. And, if Intel is the partner, I can see most of the $4 Billion of R&D budget immediately stashed. Hey, if you've got the team that did it all for less than a $1 billion a year in R&D, wouldn't it be logical that that's who you would keep.

How about Apple and Intel merging? Let's see if Paul Otellinni and Steve Jobs can get along in their new venture and whether or not, Apple may be a good grab for Intel as its Microsoft alliance is softening.

PCs Are Not Going Away – This Year

The PC Industry formed in earnest in August, 1981. 2006 marks the 25th anniversary of the PC. There are about 800 million PCs installed worldwide and there are about 125 million game consoles. Most of those PCs have Intel inside while the game consoles are split between Xbox, Playstation, and Nintendo models. There are a number of industry analysts who are looking for the game console industry to take off and for the PC industry to taper off and eventually begin to shrink. But, when?

Any of these next generation game consoles can be made to be more than game consoles. They can become bonafide PCs running Linux, IBM's OS/400 (i5/OS), Unix, and even Windows. So, if the game console can be a PC and the PC can run games, which is better? We can expect game consoles to be loaded with software like spreadsheets and word processors in the future and become the central point in homes for things like security systems or lighting control or music or lawn watering or closed circuit TV or other things that we can yet imagine. The supercomputing nature of the game consoles, which are about to become prevalent, position them to become the dominant type of computer in the home.

Of course, it will take Dad awhile to accept using an Xbox 360 for spreadsheets and document production, but Aunt Susie who likes her Web TV unit will just be tickled to no end to use her new Playstation as an email / Web browser that can actually store things on disk. The future home unit will be the game console for sure. But when? Maybe sooner than you think! It's not here yet but when it comes, and it can do as much as a PC and faster, it will take the home by storm. Then, more and more POWER or Cell chips will be needed to replace those clumsy old Pentium D chips in those old PCs. Then, Intel is in trouble. Somehow, AMD will figure something out. Maybe it's time to buy IBM stock.

Final Words on Intel

One might ask with all these IBM developments, exactly how far behind is Intel with parallelism, threading, cooling, ASIC, game console, and supercomputing work? It is bold, yet fair to ask: "Is there actually room for Intel in the coming chip world?"

There may not be. And wouldn't it be funny if it is not the governments of the world who are the final arbiters of Intel's survival. It may very well be your ten year old who makes the decision.

Index

LETS GO PUBLISH! Books:

(sold at www.itjungle.com, www.mcpressonline.com, www.iseriesnetwork.com)

LETS GO PUBLISH! www.letsgopublish.com Our titles include the following: email info@letsgopublish.com for ordering information

Getting Started With The WebSphere Development Studio for iSeries .
Your introduction to the new IBM strategy for Application Development. Includes a case study and examples of UI / Logic separation and CPW savings techniques.

The iSeries Pocket Developers' Guide.
Comprehensive Pocket Guide to all of the AS/400 and iSeries development tools - DFU, SDA, etc. You'll also get a big bonus with chapters on Architecture, Work Management, and Subfile Coding.

The iSeries Pocket Database Guide.
Complete Pocket Guide to iSeries integrated relational database (DB2/400) – physical and logical files and DB operations - Union, Projection, Join, etc. Written in a part tutorial and part reference style, this book has tons of DDS coding samples, from the simple to the sublime.

The iSeries Pocket Query Guide.
If you have been spending money for years educating your Query users, and you find you are still spending, or you've given up, this book is right for you. This one QuikCourse covers all Query options.

Getting Started With The WebSphere Development Studio Client for iSeries (WDSc) Focus on client server and the Web. Your introduction to the client server and web development tools. Includes CODE/400, VisualAge RPG, CGI, WebFacing, and WebSphere Studio. Case study continues from the Interactive Book.

The iSeries Pocket WebFacing Primer.
This book gets you started immediately with WebFacing. A sample case study is used as the basis for a conversion to WebFacing. This interactive 5250 application is WebFaced in a case study format before your eyes. Either learn by reading the book or read while working along on your own system.

Getting Started with WebSphere Express Server for iSeries: Your Step-by-Step Guide for Setting Up WAS Express Servers
A Comprehensive guide to setting up and using WebSphere Express. It is filled with examples, and structured in a tutorial fashion for easy learning. The book is designed to take you to a point at which you understand the notion of a servlet server, what WebSphere Express is, where it came from, how to order it, how to set it up, and how to make it work in your shop.

The WebFacing Application Design & Development Guide:
The Step by Step Guide to designing green screen iSeries applications for the Web. This is both a systems design guide and a developers guide. Using this guide, you will understand how to design and develop Web applications using regular workstation interactive RPG or COBOL programs. When you learn the tricks, and observe the sample code in action, you might choose to develop all your applications using this approach.

The iSeries Express Web Implementor's Guide. Your one stop guide to ordering, installing, fixing, configuring, and using WebSphere Express, Apache, WebFacing, iSeries Access for Web, and HATS/LE.

Can the AS/400 Survive IBM?
Exciting book about the AS/400 in an iSeries World.

The All-Everything Machine
The story about IBM's finest computer server.

Chip Wars
The story of the ongoing war between Intel and AMD and the upcoming was between Intel and IBM. This book may cause you to buy or sell somebody's stock.